Readings from the
Encyclopedia of Neuroscience

Sensory Systems II
Senses Other than Vision

Selected and with an Introduction by
Jeremy M. Wolfe

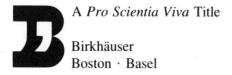
A *Pro Scientia Viva* Title

Birkhäuser
Boston · Basel

Library of Congress Cataloging-in-Publication Data
(Revised for vol. 2.)
Sensory systems.

  (Readings from the Encyclopedia of neuroscience)
  ''A pro scientia viva title.''
  Bibliography: v. 1,   v. 2, p.
  Contents: 1. Vision and visual systems/selected
and with an introduction by Richard Held—2. Senses
other than vision/selected and with an introduction
by Jeremy M. Wolfe.

  1. Senses and sensation. 2. Vision. I. Held,
Richard.
QP431.S457 1988   591.1′82   88-19364
ISBN 0-8176-3395-2 (vol. 1)
ISBN 0-8176-3396-0 (vol. 2)

CIP-Titelaufnahme der Deutschen Bibliothek
Readings from the Encyclopedia of neuroscience.—Boston ;
Basel : Birkhäuser.
  (A pro scientia viva title)
Sensory systems.
  2. Other senses.—1988
Sensory systems.—Boston ; Basel : Birkhäuser.
  (Readings from the Encyclopedia of neuroscience)
  (A pro scientia viva title)
2. Other senses / selected and with an introd. by Jeremy M.
  Wolfe.—1988
  ISBN 3-7643-3396-0 (Basel) brosch.
  ISBN 0-8176-3396-0 (Boston) brosch.
NE: Wolfe, Jeremy M. [Hrsg.]

Printed on acid-free paper.

Printed and bound by Edwards Brothers Incorporated, Ann Arbor, Michigan.
Printed in the United States of America.

9 8 7 6 5 4 3 2 1

ISBN 0-8176-3396-0
ISBN 3-7643-3396-0

Series Preface

This series of books, ''Readings from the *Encyclopedia of Neuroscience*,'' consists of collections of subject-clustered articles taken from the *Encyclopedia of Neuroscience*.

The *Encyclopedia of Neuroscience* is a reference source and compendium of more than 700 articles written by world authorities and covering all of neuroscience. We define neuroscience broadly as including all those fields that have as a primary goal the understanding of how the brain and nervous system work to mediate/control behavior, including the mental behavior of humans.

Those interested in specific aspects of the neurosciences, particular subject areas or specialties, can of course browse through the alphabetically arranged articles of the *Encyclopedia* or use its index to find the topics they wish to read. However, for those readers—students, specialists, or others—who will find it useful to have collections of subject-clustered articles from the *Encyclopedia*, we issue this series of ''Readings'' in paperback.

Students in neuroscience, psychology, medicine, biology, the mental health professions, and other disciplines will find that these collections provide concise summaries of cutting-edge research in rapidly advancing fields. The nonspecialist reader will find them useful summary statements of important neuroscience areas. Each collection was compiled, and includes an introductory essay, by an authority in that field.

George Adelman
Editor,
Encyclopedia of Neuroscience

Contents

Introduction

The Five Senses

Here we have a small book on the senses (excluding vision, which gets a companion volume of its own) made up of entries from the *Encyclopedia of Neuroscience*. One might expect to look at the Contents and see a straightforward list of those senses, but it is not that simple. How many senses are there? The traditional answer is "five": vision, hearing, taste, smell, and touch. This list goes back to Aristotle (c. 350 BC) and even he knew that there was a problem with this formulation. For Aristotle, a sense had a single "special object" that was analyzed by that sense and that sense alone. "Color is the special object of sight, sound of hearing, flavor of taste." But, he goes on to say, "Touch . . . discriminates more than one set of different qualities." By "different qualities," Aristotle is referring to sensations like hot and cold, pain, perhaps itch and tickle. Are these sensations part of a unitary sense of touch? If you look in this book, you will find separate articles on thermoreceptors, nociceptors (pain), and mechanoreceptors (touch), suggesting that the sense of touch is, in fact, a set of senses. Dividing the sense of touch is a matter of convenience and not of settled fact. At one time or another arguments have been put forward granting itch, tickle, vibration, pressure, and several other sensations standing as independent senses (Geldard, 1972). Modern texts often cautiously refer to the "skin or cutaneous senses."

Abandoning the unity of one of the classic senses calls the unity of the others into doubt. If touch can be subdivided, what about vision or hearing? Is there a sense of "red" distinct from a sense of "blue" or of "bright" or "vertical"? In the 19th century, Helmholtz proposed that a single sensory "modality" was one where all the stimuli could be connected in some continuous fashion. Thus, red and blue are not the product of two senses because there is a continuous set of stimuli between red and blue. The same cannot be said for touch and hot. The modality argument can be used to claim vision and audition as unitary senses, but taste and smell now become problematical. There are four basic tastes: sweet, sour, salty, and bitter. Though they are not continuous with one another, we do not seem to feel the need to talk about the "taste senses" perhaps because one organ seems to be responsible for all four. Just as the eye is responsible for vision and the ear for hearing, the tongue seems to be responsible for taste. This is not true in any strict sense. The tongue is merely the holder for most of the taste cells (see "Taste Bud"), but it does feel like a unitary sense organ. The basic tastes can be treated independently. An example might be the article in this book entitled "Sodium Appetite," in which saltiness is the only taste that counts.

The situation in smell is worse. After 100 years of effort, there is no agreement on a limited set of basic smells. If anything, there is some agreement that no such set exists. Each smell seems to be a unique entity. There is no clear, continuous path from, say, "minty" to "flowery." When we do attempt to impose an introspective order on olfactory stimuli, it is usually an affective order. Some smells are pleasant and some are not. This is quite different from the intuitively obvious order of colors in the spectrum or tones on a scale.

It is worth noting that, perceptually, the distinction between smell and taste is not terribly clear. When Aristotle claimed that "flavor" was the "special object" of taste, he was wrong. The flavor of a food is a combination of taste and smell with the lion's share of influence going to smell. Many flavorful substances have little or no taste. Cinnamon is a particularly vivid example. If you hold your nose closed and place cinnamon on your tongue, you will find that it has no more taste than so much dust. Open your nose and the flavor of cinnamon will fill your head. This is not to argue that taste and smell are not distinct in humans, merely to point out that the anatomical distinction is clearer than the perceptual distinction.

Beyond the Five Senses

Difficulties of classification within the five Aristotelean senses are just the beginning of the problem of the enumeration of the senses. Several sensory modalities are entirely absent from his list. For example, what is dizziness? The only likely candidate among the original five is "touch," and that does not seem correct. Dizziness and, more generally, information about gravity, body motion, and the position of the head in space are the province of the vestibular system. The vestibular apparatus is found near the inner ear. Indeed, the sensory mechanisms of hearing and the vestibular system are remarkably similar (see the articles entitled "Gravitational Effects on Brain and Behavior," "Vestibular System," and "Visual-Vestibular Interaction").

This problem has been evident for a long time, leading to various efforts to amend the basic list. In his discussion of the senses, James Mill (1829) added sensations of disorganization (which included pain), muscular sensations, and sensations of the alimentary canal to the basic five. Alexander Bain talked about the sensations of "organic life." These additions refer to the collection of sensations that inform us about the position and disposition of other parts of our bodies. Our muscles and joints tell us about the position of our limbs (see "Muscle Receptors, Mammalian" and "Muscle Sense"). They also report on their state of rest or fatigue. Signals from deep inside our bodies: aches and pains and other less objectionable sensations seem to share something with the skin senses but are, by definition, not skin senses. You will see, if you turn to the entry entitled "Thermoreceptors," that different types of temperature sensors exist in the skin, in the gut, and even in the brain. Activity in the internal receptors has only limited perceptual consequence but is important in the regulation of body temperature. Sensations from mechanoreceptors and nociceptors deep in the body can accompany overindulgence at the dinner table or an encounter with the flu. Titchener (1896) included sensations from the lungs, the blood vessels (with a question mark), and the sex organs. The last of these might be grouped under the skin senses but is probably worth giving independent status (see "The Genital Sensory System"). In his fine book on the senses, Geldard (1972) includes a list of sense-qualities that have at one time or another been proposed as having independent status: "pressure, contact, deep pressure, prick pain, deep pain, quick pain, warmth, cold, heat, muscular pressure, articular pressure, tendinous strain, ampullar sensation or dizziness, vestibular sensation or sense of translation, appetite, hunger, thirst, nausea, sex, cardiac sensation . . . pulmonary sensation . . . itch, tickle, vibration, suffocation, satiety, and repletion" (p. 258) and, you will note, that list does not include possibilities from smell, taste, vision, or hearing.

The Senses of Animals

Matters become still more complicated when we turn to animals. Their senses can be organized differently from ours and they can possess senses that we lack altogether. For example, in humans the distinction between smell and taste is reasonably clear. Smell takes place in the nose and taste, on and around the tongue. What shall we say about animals like the fly, whose receptors for sweet molecules are on its feet, or the catfish, whose "taste buds" cover the entire body surface? Are they smelling or are they tasting as they step in or swim through solutions? In this volume, the problem is made clear in an entry like "Chemotaxis, Bacterial: A Model for Sensory Receptor Systems." Bacteria clearly respond to chemical stimuli. Are they smelling or tasting? To avoid the need to make a distinction, smell and taste are often grouped together under the heading of chemical senses.

Within sensory systems that we have in common, animals can have sensitivities that we do not have. Whales can hear sound frequencies lower than we can hear. Any number of animals can hear at higher frequencies (e.g., dogs, bats, and insects). Animals can use senses in ways that we cannot. The ability of bats to navigate by echolocation is an example (''Echolocation''). Finally, there are senses that we lack altogether. The electrosensory system allows some fish to navigate in murky water by generating an electrical field and then sensing distortions in that field produced by surrounding material (''Electric Organs, Fishes''). When the internal generator is big enough, it is possible to use the electric pulse to stun prey. However, most of the electric fishes and electric eels generate electricity for sensory and not for aggressive purposes. The lateral line organ is another system used by fish to navigate and is related both to touch and to the vestibular senses (''Lateral Line System''). Birds and a number of other species can navigate by sensing the earth's magnetic field (''Bird Navigation''). While some workers believe there is a magnetic sense in humans (Baker, 1981), others have their doubts (Gould and Able, 1981). Many animals have a supplementary olfactory system, the vomeronasal system. Again, evidence for its presence in humans is weak.

In sum, if there is a source of useful information in the environment, from polarization of sunlight (again, ''Bird Navigation'') to chemical communications between insects (''Pheromones''), some animals will have developed a sensory apparatus to take advantage of that information. The result is an array of senses far richer than the five described by Aristotle. The collection of articles in this volume are a valuable entry point into the realm of the senses. The references that accompany each article can lead the reader into the details of any given topic. With that introduction, the precise enumeration of the senses is left to you, the reader. Let me know if you have the answer.

Since we cannot provide a simple list of the senses, we have opted to place the articles in this collection in alphabetical order. To make the collection more useful, we offer the following ''second Contents,'' organized by topic.

References

Aristotle (c. 350 BC): *De Anima*, Book 2, chapters 5-12 (reprinted in many anthologies, e.g., Herrnstein and Boring, 1965)

Bain, A (1856): *The Senses and the Intellect*

Baker, R R (1981): *Human Navigation and the Sixth Sense*. New York: Simon and Schuster

Boring, E B (1942): *Sensation and Perception in the History of Psychology*. New York: Appleton, Century, Crofts

Geldard, F A (1972): *The Human Senses*, 2nd Ed., New York: John Wiley & Sons.

Gould J L and Able K P (1981): Human homing: An elusive phenomenon. *Science* 212: 1061–1063

Herrnstein R J and Boring E B (1965): *A Source Book in the History of Psychology*. Cambridge, MA: Harvard U. Press

Mill, James (1829): *Analysis of the Phenomena of the Human Mind*

Titchener, E B (1896): *An Outline of Psychology*. New York: Macmillan

Jeremy M. Wolfe
Department of Brain and Cognitive Sciences
Massachusetts Institute of Technology

Alliesthesia

Michel Cabanac

The word alliesthesia (Greek *Allios* changed and *-esthesia* sensation) is applied to the affective component of sensation, pleasure or displeasure. The amount of pleasure or displeasure aroused by a given stimulus is not invariable—it depends on the internal state of the stimulated subject. Factors that can modify the internal state and in turn induce alliesthesia are as follows: internal physiological variables (e.g., deep body temperature or body dehydration modify the pleasure of thermal sensation or taste of water); set points (e.g., during fever the body temperature set point is raised and pleasure defends the elevated set point); multiple peripheral stimuli (e.g., mean skin temperature determines the set point for deep body temperature and in turn generates alliesthesia); and past history of the subject (e.g., association of a flavor with a disease or a recovery from disease renders it unpleasant or pleasant). Positive alliesthesia indicates a change to a more pleasurable sensation; negative alliesthesia a change to a less pleasurable one.

Further reading

Attia M (1985): Thermal pleasantness and temperature regulation in man. *Neurosci Biobehav Rev* 8:335–342

Cabanac M (1971): Physiological role of pleasure. *Science* 173:1103–1107

Fantino M (1985): Role of sensory input in the control of food intake. *J Auton Nerv Syst* 10:347–359

Auditory Prosthesis

Gerald E. Loeb

Functional auditory prostheses provide an information-carrying interface between electronic systems and neuronal systems. They provide coherent sensations of sound in patients suffering from profound (usually total) deafness as a result of various traumatic, infectious, and hereditary diseases that cause loss of the cochlear hair cells. These cells normally transduce the mechanical vibrations of sound waves (entering the fluid-filled cochlear chambers, via the middle ear ossicular chain) into bioelectric impulses to the brain. The axons of the spiral ganglion cells which make up the auditory nerve convey these impulses from the organ of Corti on the basilar membrane (where the hair cells are located) to the brain stem auditory nuclei. In the absence of the hair cells, the remaining spiral ganglion cells can be activated directly by electrical currents

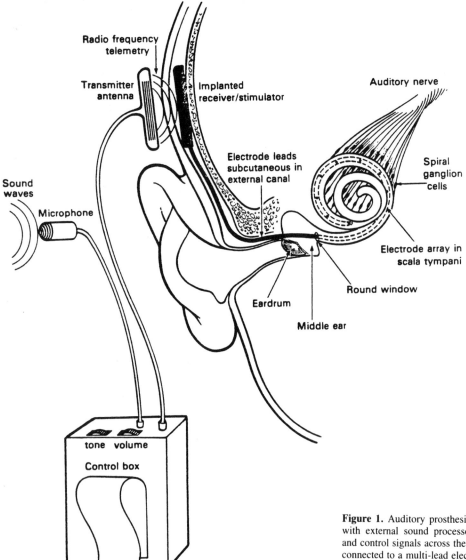

Figure 1. Auditory prosthesis. Typical multichannel cochlear prosthesis with external sound processor, radio frequency transmission of power and control signals across the skin, and implanted receiver and stimulator connected to a multi-lead electrode array inserted 1½ turns into the cochlear spiral.

from stimulating electrodes either in or near the cochlear chambers (usually at the round window entrance or in the scala tympani) or in the auditory nerve proper as it courses through the bony modiolar canal. The auditory nerve activity so elicited is interpreted as sound by the central nervous system, with loudness, pitch, and timbre apparently dependent on the amplitude, location, and waveform of the stimulating current.

About a dozen different single-channel and multichannel devices are now in various stages of animal and clinical research and are beginning to be commercially available. All share the general components shown in Figure 1, including external parts similar to a hearing aid (battery-powered, wearable control box with a microphone) plus a radio telemetry link that conveys information and power to the implanted electronics for generating the stimulus waveform and conveying it to the desired location(s).

Single-channel devices provide only a single, temporally modulated waveform to a single stimulating electrode site. This signal may be simply a filtered and dynamically compressed version of the acoustic signal picked up by the microphone or it may be synthesized by complex algorithms for preprocessing the acoustic signal to enhance the intelligibility of human speech. Regardless of the form of processing or the site of the stimulating electrode, most patients appear to obtain similar benefits from single-channel prostheses, including significantly enhanced lip-reading, improved modulation of their own voices, and useful awareness and identification of ambient sounds. Both theoretical considerations and empirical results suggest that unaided speech perception probably is not possible with single-channel prostheses.

The approaches and results of multichannel prostheses, most of which are still in early stages of clinical research, are much less easily summarized. All the multichannel systems attempt to make use of the tonotopic organization of the normal cochlea, in which high-frequency sounds are normally transduced at the basal end of the mechanically tuned basilar membrane, and lower frequencies are represented progressively more apically in the spiral. Multicontact electrode arrays have been implanted in the cochlea so as to permit local stimulation of subsets of the spiral ganglion cells. As might be expected, the stimulation of several such adjacent sites individually gives rise to sensations of sound with a distinct spatial ordering of pitch, at least as long as the stimulus intensity is low enough to activate only those ganglion cells in the immediate vicinity of each stimulating electrode. Unfortunately, the quality of the sound is not at all tonelike, even for presumably well-localized stimulation, regardless of the frequency or waveform of the electrical stimulus.

Despite this limitation, most multichannel prostheses have proceeded with stimulation schemes based on the principle of a frequency-channel vocoder synthesizer. Intelligible speech can be synthetically reproduced by six to eight single-frequency sound sources, each temporally modulated in proportion to the instantaneous power present in the band of frequencies immediately surrounding the center frequency of the source oscillator. Multichannel prostheses usually filter the incoming acoustic information into bands corresponding to the place-pitch sensations evoked by each available stimulating electrode site, and they use multichannel telemetry to modulate appropriately the electrical activation of each site. Systems with just four such independent channels have resulted in significant levels of word recognition in some but not all deaf subjects. As many as eight independent channels are probably feasible based on biophysical and technological considerations.

The ultimate capabilities of multichannel auditory prostheses will depend on progress in three key areas: 1. Biomaterials engineering. With the technology and materials currently available, we have probably gone as far as we can in the fabrication of complex devices and their safe surgical insertion and long-term operation in the delicate and confined space of the cochlea. 2. Electrophysiological factors. The selective activation and external control of one group of neurons, located in electrically conductive fluids and adjacent to other neurons, is a complex biophysical problem. Critical factors include the exact location, size, and orientation of the electrode contacts as well as the numbers and conditions of the remaining spiral ganglion cells; the latter appear to be highly variable among deaf subjects. 3. Speech processing. It is not clear why or how the various electrical stimulation parameters give rise to the particular sound sensations that have been reported, nor is there any theory to suggest how to transform optimally the acoustic speech signal into electrical stimuli that will provide the critical cues for word recognition despite the complex distortions.

Further reading

Parkins CW, Anderson SW, eds. (1983): Cochlear prosthesis: An international symposium. *Ann NY Acad Sci* 405:1–532

Schindler RA, Merzenich MM, eds. (1985): *Cochlear Implants*. New York: Raven Press

Auditory System

John F. Brugge

The auditory system is remarkable in the range of sound frequencies and intensities it can detect and in the small differences in these parameters it can discriminate. A young listener can hear sounds ranging in frequency from 20 Hz (cycles per second) to 20,000 Hz (20kHz). Within this range, as little as a 0.1% change in frequency is detectable. In the intensity domain, the same listener detects displacements of the ear drum two orders of magnitude smaller than the diameter of a hydrogen atom. At the same time, hearing is quite clear when the amplitude of the sound is raised by a factor of 10^6, which gives good listeners a dynamic range of more than 100 decibels (dB) on the scale of acoustic energy. Within this dynamic range, a change of 1 or 2 dB is easily detected. Our listener may detect with uncanny accuracy the location of a sound in space and will discriminate between two speakers located within a few degrees of each other on the horizontal plane. This ability to both detect and discriminate sounds is achieved by mechanisms operating at the levels of the external ear, middle ear, and inner ear (Fig. 1) and in the central auditory pathways of the brain (Fig. 2).

External ear

The external ear consists of the pinna (auricle) and external ear canal which is closed at its central end by the ear drum or tympanic membrane. The shapes of the pinna and ear canal help to amplify sound by as much as 20 dB over a broad range of frequencies centered around 4 kHz. Pinnae vary considerably in size and shape from one mammalian species to the next and, especially in those mammals with mobile pinnae, this structure may act as a directional amplifier to aid in localizing the source of a sound in space.

Middle ear

The middle ear cavity, located just behind the tympanic membrane, is normally air-filled. The Eustachian tube connects the middle ear cavity with the nasopharynx, and the periodic opening of this tube during swallowing, yawning, chewing, etc. insures that a static atmospheric pressure is maintained in the middle ear. Airborne sound waves that enter the ear

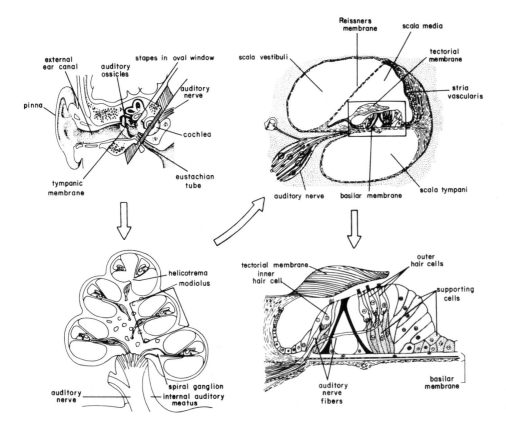

Figure 1. The structures of the external, middle, and inner ears.

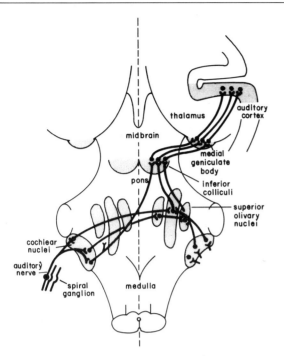

Figure 2. A schematic representation of the ascending auditory pathways of the brain.

canal strike the tympanic membrane and set it into motion. Three auditory ossicles (the malleus, incus, and stapes) transmit this motion to the inner ear. The need for such an arrangement arises because of the loss of some 99.9% of the energy of a sound wave in air that would occur should that wave encounter a boundary with a fluid, such as the fluid that fills the inner ear. This noticeable loss is equivalent to about 30 dB. Two mechanisms are involved in overcoming this mismatch in impedance between air and fluid. First, the effective area of the tympanic membrane is about 15 times that of the stapes foot plate which exerts pressure on the fluid of the inner ear. Thus, the force acting on the ear drum is concentrated through the ossicles onto a small area of the stapes footplate resulting in a pressure increase proportional to the ratio of the areas of the two structures. Second, the lever arm of the malleus is longer than that of the incus with which it articulates giving an additional mechanical advantage of about 1.3. The two factors together compensate for the theoretical loss which would occur at the air-fluid interface.

Inner ear

The inner ear, located in the temporal bone of the skull, contains the receptor organ of hearing (organ of Corti) as well as the receptors that detect head motion and position. The auditory portion, the cochlea, in mammals is a coiled and tapered fluid-filled chamber that is divided along almost its entire length by a partition. The two spaces thus formed, the scala vestibuli and scala tympani, are filled with a fluid called perilymph. They communicate with each other through an opening at the top (apex) of the cochlea called the helicotrema. At the base of the cochlea each scala terminates at a membrane (window) which faces the middle ear activity. The scala vestibuli ends at the oval window into which the footplate of the stapes rocks when the ear drum moves; the scala tympani ends at the round window, which provides a pressure relief

for movement of the cochlea fluid. The partition that divides the cochlea lengthwise is itself a fluid-filled tube called the scala media or cochlear duct. Its fluid, endolymph, is chemically different from perilymph. The cochlear duct is bounded on three sides by a bed of capillaries and secretory cells (the stria vascularis), a layer of simple squamous epithelial cells (Reissner's membrane), and the basilar membrane upon which rests the organ of Corti.

The organ of Corti consists of receptor cells (hair cells) held in place by supporting cells. Hair cells are modified epithelial cells having hairs (stereocilia) protruding from their apical ends which come close to or in contact with an overlying auxiliary structure called the tectorial membrane. The base of the hair cell is in synaptic contact with the distal ends of auditory nerve axons. Sound waves that reach the inner ear set the basilar membrane, and hence the organ of Corti, into motion. This causes a shearing motion between the tectorial membrane and the tops of the hair cells which, in turn, displaces the stereocilia. This bending of the sensory hairs initiates a chain of electrical events that leads to release of a chemical neurotransmitter at the base of the receptor cell and the generation of action potentials in the auditory nerve.

The basilar membrane varies systematically in width, and hence in stiffness, from cochlear base to apex. It is wider and more flaccid near the base than at the apex. Consequently, the amplitude of vibration at any given place along the basilar membrane is a function of the frequency of the sound. When very high frequency sound waves reach the ear, only the region nearest the base is set into motion. As the frequency of the sound is lowered, the place of maximal amplitude of vibration shifts toward the cochlear apex. Because of this gradient in frequency sensitivity, the basilar membrane is said to be tonotopically organized. The motion of the basilar membrane is one of a travelling wave of deformation which begins at the cochlear base and moves apically toward a frequency-dependent place of maximal amplitude.

The organ of Corti has two kinds of hair cells—outer and inner—which can be distinguished from one another by their location, morphology and connections with the auditory nerve. About 95% of the afferent axons of the auditory nerve are derived exclusively from inner hair cells and hence they carry the bulk of information from the inner ear to the brain. Outer hair cells are contacted by the remaining 5% of the afferent fibers but they receive the greatest proportion of axonal terminals from neurons that lie in the brain. The functions of the efferent projection system, which originates in cell groups of the brainstem, are poorly understood.

Each auditory nerve contains between 24,000 and 50,000 axons, depending on the animal species. These axons transmit to the brain trains of nerve impulses that encode all of the information needed by the central nervous system to perceive sounds in the environment. There are several mechanisms involved at the level of the cochlea and auditory nerve for encoding the frequency and intensity of a sound. Coding of more complex acoustic parameters that require the use of two ears, such as sound location or motion, takes place in the central nervous system.

Auditory nerve fibers respond within a limited domain of frequencies and intensities. This domain is called the response area of the fiber. The frequency to which the fiber is most sensitive (i.e. has the lowest threshold) is called the "best" or "characteristic" frequency (CF) of that fiber. Because of the mechanical tuning properties of the basilar membrane and the fact that a single auditory nerve fiber usually contacts but one inner hair cell, the CF can be assigned a place along the basilar membrane to which the fiber is attached. By making such a frequency-to-place transformation a distance map of the cochlea can be drawn. To the first approximation, linear cochlear distance is mapped onto a logarithmic scale of frequency. The fact that auditory neurons are frequency selective has been taken as evidence to support the "place theory" of hearing, which states that the pitch of a sound is determined by which fibers in the auditory nerve array are excited by that sound.

A second way that frequency information is transmitted in auditory nerve fibers is via a temporal code. Action potentials evoked by sounds with frequencies below about 4 kHz occur at preferred times that are integral multiples of the period of the stimulus waveform. This is a consequence of the fact that these fibers discharge in response to unilateral movement of the basilar membrane and, hence, the discharges are phase-locked to the halfwave-rectified stimulus waveform. Phase locking in auditory nerve fibers is predicted by the "volley theory" of hearing, which states that the pitch of a sound is determined by the temporal rhythm of the discharges of auditory nerve fibers. Such a temporal code seems essential for carrying the low-frequency information in human speech (e.g. the vowel sounds) and for carrying the temporal information used as a cue by listeners for localizing the source of a sound in space. Neither the place theory nor the volley theory alone fully account for a listener's ability to distinguish one tone from another and, thus, it is necessary to postulate for this a dual theory of pitch perception which incorporates both mechanisms.

Sound intensity is likely encoded, in part, by the rate at which a nerve fiber discharges. As stimulus intensity is raised, the number of discharges steadily increases up to a plateau some 30–40 dB above threshold. A second mechanism involves the recruitment of additional fibers into the active population. Both mechanisms seem necessary to achieve the nearly 100 dB of dynamic range demonstrated by human listeners.

Central auditory pathways

Auditory nerve fibers enter the brainstem, bifurcate in an orderly fashion, and terminate in a structural complex in the lateral part of the medulla known as the cochlear nuclei (CN). The topographic distribution of auditory-nerve terminals in the CN reflects that of the nerve array innervating the organ of Corti. Hence, within the CN the cochlea is represented in an orderly way and the CFs of neurons are mapped out in a "tonotopic" fashion. Tonotopy is preserved at all levels of the central auditory pathway, including the auditory cortex. Information transmitted by auditory nerve fibers is received, transformed and eventually transmitted by CN neurons to higher centers of the brainstem including the nuclei of the superior olivary complex (SOC), lateral lemnisci (NLL), and inferior colliculi (IC). Many fibers leaving the CN's on the right and left sides of the brain converge directly upon neurons of the SOC and indirectly on cells of the NLL and IC. Many of the brainstem neurons receiving such a binaural input are, in turn, sensitive to the interaural cues of time and intensity used by listeners for localizing the source of a sound in space.

Information transmitted through the brainstem reaches the medial geniculate body (MGB) of the thalamus from which it is relayed to auditory areas of the cerebral cortex. Auditory cortex comprises a tonotopically organized primary field, called area AI, and several surrounding fields all of which receive input from one or more of the subdivisions of the MGB. These cortical fields are connected with each other and with their counterparts on the opposite cerebral hemisphere via a system of highly organized cortico-cortical circuits. Cortico-thalamic and cortico-tectal projection systems provide feedback to thalamic and midbrain levels. Cortical neurons having the same CF are arrayed in columns perpendicular to the cortical surface thus giving a columnar organization to auditory cortex related to cochlear place. Binaural interactions that take place at brainstem levels are transmitted to the cortex. Cortical neurons sensitive to interaural time or intensity differences detect sounds that tend to lie in the contralateral auditory hemifield. Cells with similar binaural response properties are also grouped together and arrayed in a columnar fashion. Unlike the frequency map which reflects the spatial array of auditory nerve fibers within the cochlea, the binaural cortical map is a computational one that depends on spatial and temporal interactions evoked by stimulation of the two ears.

Further reading

Green DM (1976): *An Introduction to Hearing*. New York: Wiley
Moore BCJ (1977): *Introduction to the Psychology of Hearing*. Baltimore: University Park Press
Yost WA, Nielsen DW (1977): *Fundamentals of Hearing*. An Introduction. New York: Holt, Rinehart and Winston

Barrels, Vibrissae, and Topographic Representations

Thomas A. Woolsey

In the cerebral cortex, a barrel is a group of neurons shaped as such and visible with a microscope and in some cases with the naked eye. Each barrel is a cylindroid located in the middle layers of the somatosensory cortex of many rodents and certain other mammals and receives segregated inputs related to individual tactile organs called vibrissae or whiskers located principally on the face (see Fig. 1). A vibrissa consists of a central hair in a mechanically isolated hair follicle receiving a segregated and substantial sensory innervation known to be related to four different receptor specializations. Sensory inputs from the vibrissae can be modulated through motor control. Striated muscles actively move many of the whiskers back and forth during exploratory whisking behavior at frequencies of 7 Hz in rats and 15 Hz in mice. Each hair follicle is surrounded by erectile tissue, presumably under opponent autonomic control, which could dynamically alter the functional threshold of the tactile organs. The vibrissae are arranged in species-specific, stereotyped, grid-like patterns in which each vibrissa has a unique position.

Collectively, the cortical barrels define an architectonic field—the barrel field. When viewed as a single plane, the barrel field is homeomorphic to the whisker pattern on the face. Within the barrel field, each barrel has a unique position corresponding to the vibrissa projecting to it. There is a precise topographic anatomical correlation between the sensory periphery and the barrel cortex. The relationship has been verified in three independent ways: (1) Microelectrode recordings show that most cells in a particular barrel are activated by the appropriate contralateral whisker. (2) Metabolic markers which measure short- and long-term changes in activity are altered in corresponding barrels after manipulation of selected vibrissae.

(3) Selective surgical manipulation of the vibrissal nerves in the early postnatal period alters the development of the corresponding barrels. In a number of animals it has been clearly demonstrated that the densities of layer IV neurons when viewed in single sections parallel to the cortical surface describe an outline of the animal's face, body, legs, and tail. This visible representation of the body in the somatosensory cortex is a direct example of the general principle of topographic representation of sensory and motor functions in the brain which have long been known from clinical observations and correlative neuropathology, electrophysiological recording, and experimental neuroanatomy. The principle of topographic representation states that many central nervous system structures contain a systematic, topographically correct representation of sensory inputs from the body, eyes, and ears and to motor outputs for voluntary movement.

The direct pathway from the vibrissae to the barrels crosses three synapses (see Fig. 2). Cell bodies of the afferent fibers are located in the trigeminal ganglion, and the central processes of these neurons terminate in three separate ipsilateral brain stem nuclei. Cells from the rostral two nuclei project to the contralateral somatosensory thalamus. Neurons in the thalamus project to the somatosensory cortex of the same side. A characteristic of the pathway is that the afferent terminals in each of these loci are grouped in whisker-like patterns, as are the target cells. The patterns are consistent from animal to animal and can be demonstrated in sections cut in specific planes with and without routine staining methods. The neurons of the trigeminal ganglion are the only ones in the pathway which do not exhibit a precise whisker-related pattern.

Much is known about the anatomy and function of the cere-

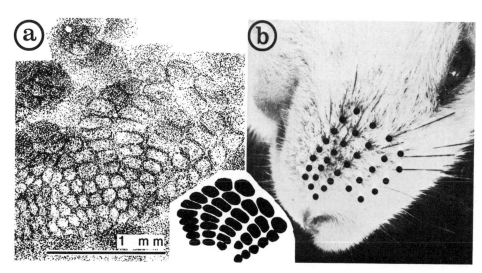

Figure 1. Relation of cortical barrels to individual hairs on the snout of rodents. A. Photomicrograph of a Nissl-stained tangential section through the relevant region of somatosensory cortex. Each barrel is outlined by a ring of darkly staining neurons. B. Photograph of the right face of a mouse showing the mystacial vibrissae; the position of each major hair is indicated by a black dot (axons from the hairs on the right face give rise to pathways projecting to the left cortex). Each vibrissa corresponds to an individual barrel. The inset shows the overall arrangement of barrels in the cortex.

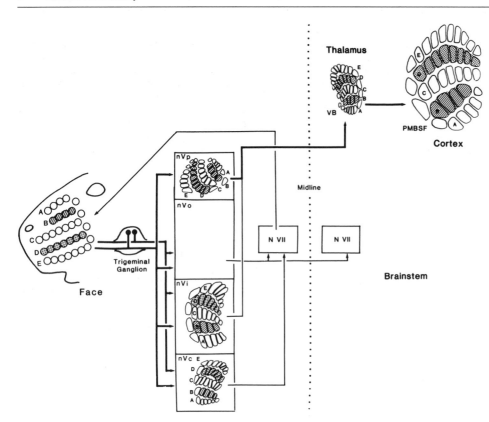

Figure 2. Schematic diagram summarizing the principal ascending connections in the rodent trigeminal pathway. Somatotopically organized projections which are isomorphic to the pattern of the whiskers on the face can be demonstrated in the brain stem trigeminal complex (Sub-nuclei caudalis–nVc and interpolaris–nVi of the spinal trigeminal nucleus and the principal sensory nucleus of V–nVp), the thalamus (VB), and the posteromedial barrel subfield of the barrel cortex (PMBSF).

bral cortex because of its accessibility, planar geometry, and relative size. The barrels are located in cortical layer IV. The biggest barrels in mice are ovoid with a major diameter of about 300 μm. Cells are arranged in rings around dense clusters of afferent endings from the thalamus. There are about 2500 neurons in the largest barrels. The number of these neurons in the tangential cross section of a barrel is directly proportional to the number of fibers innervating the appropriate contralateral whisker. The axons and dendrites of layer IV stellate cells are largely confined to the barrel in which their soma is located. The stellate cells are of two principal types—spiny and smooth—which are approximately equal in number. Some, if not all, of the smooth cells are positive for glutamic acid decarboxylase (GAD)-like immunoreactivity. About 90% of the synapses in a barrel are asymmetrical (excitatory) and of these about 25% are from the somatosensory thalamus. Both stellate cells receive thalamic synapses, but those on the smooth cells are distributed on the soma and proximal dendrites while those on the spiny cells terminate on spines at regular intervals.

As in other cortices, especially sensory cortices, the predominant organization of the barrel cortex is vertical, or columnar. Whisker-related segregated thalamic projections are largely to layer IV but are present also in upper layer VI. From Golgi materials the axonal and dendritic distributions of neurons in the barrel cortex are principally interlaminar. Deoxyglucose labeling after stimulation of a single whisker labels a single column of cells and neuropil which coincides precisely win the tangential dimension with the boundaries of the relevant barrel and extends throughout the full cortical thickness. The response properties of neurons in layer IV, and to a lesser extent those in layer VI, closely resemble those described for the primary afferents. These cells respond principally to deflections of a single vibrissa, have preferences for whisker

movements in particular directions, have sustained or transient discharges to whisker movements, can code for amplitude and rate of displacement, and may follow sinusoidal stimuli to frequencies exceeding 100 Hz. Cells in other cortical layers have more specific stimulus requirements. They have precise directional responses, often are inhibited by certain forms of stimulation, and integrate information from several different vibrissae. Recent studies have identified cells in layer V which are responsive to particular spatial and temporal sequences of movement of adjacent whiskers as if the neurons were detectors for objects of finite sizes touched by the whisker hairs. Thus the anatomical and functional organization of the barrel cortex resembles in many ways the characteristics which have been so well described in the primate visual cortex, including a correspondence between the absolute dimensions of the cortical columns.

For studies in mammalian neural development, the whisker-barrel system has a number of advantages. Some rodents such as hamsters, mice, and rats are born in a very immature state. Although the neurons of this pathway are generated *in utero*, at birth cortical cells are still migrating, the segregation of the central axonal projections along the pathway is not yet complete, and most of the target neurons in the brain are only beginning to differentiate. The stereotyped and punctate pattern of the vibrissae is recognizable at birth, and selected lesions can be made in the periphery which will alter the development of the central nervous system in largely predictable and time-dependent ways. When peripheral lesions are made in different patterns at different postnatal times, the changes in the architecture and function of the central nervous system suggest a number of generalizations which are likely to apply to all pathways in the central nervous system. The pattern of central projections can be instructed from the periphery and transmitted from one central locus to the next. The pattern is related to relative rather than absolute innervation densities in the periphery. There are temporal gradients in development which are directly related to the time of neurogenesis along the neuraxis as well as within a particular part of the brain. Experimentally altered central projection patterns nonetheless have appropriate functional characteristics even though their absolute dimensions may have changed.

Sensory deprivation does not appear to alter the patterns of central projections. Deprivation can be done by simply cutting the whisker hairs in neonates and adults and can be reversed by letting the whisker hairs grow back. Preliminary findings suggest that the functional characteristics of cortical neurons are less specific in deprived animals after neonatal whisker clipping than in normal animals. In adults it has been shown that whisker clipping changes a number of biochemical parameters related to energy metabolism.

Mice are the mammalian species of choice for studies of genetics. Strains of mice with various abnormalities of the central nervous system and vibrissal patterns have been described and some have known genetics. These animals could be used to understand better the molecular basis of central nervous system development and peripheral pattern specification. An outstanding problem is the mechanism by which the organization of the periphery is reflected in the organization of the central nervous system. Because of the patterned organization of the system, interactions between central axons and their targets in development can be examined in detail. The arrangement of the whisker-barrel system means that details of central sensory processing can be approached directly with attention to the anatomical and pharmacologic organization of cortical neuronal assemblies. Finally the complex yet stereotyped patterns of exploratory behavior in which the animals use their whiskers can be investigated from a variety of viewpoints including the anatomy and the function of brain stem circuits, the psychophysical analysis of information available to behaving animals from the whiskers, and indeed the role of the whiskers in establishing social heirarchies.

Further reading

Woolsey C (1964): Cortical localization as defined by evoked potential and electrical stimulation studies. In: *Cerebral Localization and Organization*, Schaltenbrand G, Woolsey C, eds. Madison: University of Wisconsin Press

Woolsey T, Durham D, Harris R, Simons D, Valentino K (1981): Somatosensory development. In: *Development of Perception: Psychobiological Perspectives*, Aslin R, Alberts J, Peterson M, eds. New York: Academic Press

Woolsey T (1984): The postnatal development and plasticity of the somatosensory system. In: *Neuronal Growth and Plasticity*, Kuno M, ed. Tokyo: Japan Scientific Societies Press

Bird Navigation

Charles Walcott

Both the annual migrations of birds between their wintering and breeding grounds and their ability to return home when displaced from their nest involve navigation. Bird banding studies show that many migratory birds return to the same nest sites year after year. This ability implies that they can locate a specific area, not simply fly north or south with the seasons.

In the early 1950s, Donald Griffin distinguished three classes of orientation: type I was piloting or orientation by using familiar landmarks; type II was orientation in a compass direction; and type III was true navigation, in which a bird selects the home direction when released in unfamiliar territory. A migratory or homing bird probably uses all three of these types. In unfamiliar territory it might use navigation to select a homeward direction, a compass to fly in that direction, and then familiar landmarks as it nears its goal. In fact that description is probably close to the truth for homing pigeons and for many migrating birds. The real issue is: On what sensory cues are each of these types of orientation based?

Landmarks are usually thought of as physical objects such as prominent rivers, buildings, mountains, or other physiographic features. Yet they could equally well be prominent magnetic features, odor sources, low-frequency sound sources, or the like. To what extent migratory birds rely on such cues during their trips is not known; but there is some evidence that birds may respond to leading lines like prominent rivers to correct for wind drift. Yet homing pigeons equipped with frosted lenses that severely restrict their visual acuity return to the vicinity of the loft and some actually land on the loft itself! This result plus the fact that birds often migrate at night or on top of clouds when visual landmarks would not be available suggests that piloting alone cannot explain what birds do.

Birds use several cues as compasses. Day migrating birds and homing pigeons use the sun. Exposing such birds to a day/night cycle out of phase with the real day, a process known as clock-shifting, results in a compass error corresponding to the amount of shift. A six-hour shift deflects a pigeon by about 90 degrees; just what one would predict if the pigeon were compensating for the sun's apparent movement.

Indigo buntings appear to use the stars as a compass. Caged buntings in a planetarium use stars located within 30 degrees of the Pole star to find north. Young buntings learn the pattern from observing the axis of celestial rotation during their first summer. But there are many occasions when neither sun nor stars would be visible to a migrating bird; yet radar data shows that birds are well oriented even under overcast conditions. Wiltschko and colleagues showed that European robins and several species of warblers can orient in cages in the absence of celestial cues. The direction of this orientation is related to the local magnetic field; reversing the direction of the field reversed the orientation. For homing pigeons small bar magnets attached to the pigeons backs had no effect when the sun was visible, but under overcast skies these pigeons were often disoriented. Putting small coils that induced an earth-strength field around the pigeons' heads also had no effect under sunny skies; however, under overcast skies, birds with the applied magnetic field oriented in one direction homed normally. But pigeons with the opposite polarity flew 180 degrees away from home. The pigeons homed when the batteries died. It appears that under overcast skies migrating birds can use the earth's magnetic field as an auxiliary compass.

Recently it has been shown that there is an interaction between the several compass systems. Homing pigeons may rely on the magnetic compass to calibrate the sun compass, and migratory warblers may use the magnetic compass to calibrate the star compass or even the reverse. Extensive as this list of compasses may be, it does not exhaust the possibilities. Night migrants use the direction of sunset glow as a directional cue. White-throated sparrows exposed to the sunset sky through a polarizing filter oriented in a direction that depended upon the alignment of the filter. Finally, the wind itself may be important. Migratory birds in the southeastern United States often fly downwind even if such behavior takes them away from their goal. This behavior is less common elsewhere.

In contrast to the plethora of compasses, we do not have even one navigational system that is fully acceptable. Indeed, it has not yet been clearly established that migratory birds navigate at all, although it seems highly likely that they can correct their courses when drifted by the wind. Some species have been shown to home from what is presumably unfamiliar territory. Most investigations of true navigation have therefore been conducted with homing pigeons. And for pigeons there are two major suggestions for potential cues: odor and the earth's magnetic field.

Floriano Papi and colleagues at the University of Pisa, Italy, have proposed that homing pigeons use an olfactory map. Birds growing up in a loft learn to relate wind-borne odors to the direction from which the wind blows. This information is combined with learning the sequence of odors on the way to the release point. Three types of experiments support this idea: First, interference with the pigeon's olfaction leads to disorientation and little or no homing. Second, the manipulation of olfactory cues during the trip to the release site leads to differences in orientation. Third, distorting the learning of appropriate wind/olfactory information at home also leads to a predictable deflection. Difficulties in repeating many of these exciting and interesting experiments in other parts of the world lead some investigators to suspect that olfaction may not be the only basis of the pigeon's navigation.

A second possibility is that pigeons use the earth's magnetic field not just as an auxiliary compass but as part of their map as well. The earth's magnetic field varies both in intensity and direction in a generally regular manner. Theoretically, if a bird were sufficiently sensitive to changes in the magnetic field, it might be able to use the field to find its way. Supporting evidence comes from two observations: pigeons released repeatedly at the same place choose bearings that differ from home and vary day by day. These daily variations are significantly related to the temporal irregularity of the earth's magnetic field. Second, pigeons released under sunny skies at places where the earth's magnetic field is locally irregular are disoriented. Yet the magnetic perturbations caused by both the solar wind and the magnetic anomalies are too small to have a significant effect on a magnetic compass. But magnetic cues seem unlikely to be the whole answer either. Although the magnetic field of the earth as a whole varies in a regular way, locally there is considerable irregularity. It is hard to see how an organism could detect the general pattern that it would have to use for navigation through all the local noise.

There are other possible cues that birds might use; pigeons are sensitive to sounds of as low a frequency as 0.1 Hz. Such infrasounds travel long distances and might serve as a naviga-tional aid. In addition, pigeons are sensitive to tiny changes in air pressure, to polarized light, and to ultraviolet and have an impressive memory for visual features. Whether any of these abilities play a role in the navigational process is not known. Finally, it is important to remember that the navigational system, like the compass, may depend on several alternate cues.

Further reading

Able K (1980): Mechanisms of Orientation, Navigation and Homing. In: *Animal Migration, Orientation and Navigation*, Gauthreaux S, ed. New York: Academic Press

Baker R (1984): *Bird Navigation: The Solution of a Mystery?* New York: Holmes and Meier

Baldaccini N (1983): Homing in pigeons. *Comp Biochem Physiol* 76A:639–750

Walcott C, Lednor A (1983): Bird navigation. In: *Perspectives in Ornithology*. Brush A, Clark G, eds. New York: Cambridge University Press

Chemotaxis, Bacterial: A Model for Sensory Receptor Systems

Gerald L. Hazelbauer

Like eukaryotic cells that respond to hormones, growth factors, or neurotransmitters, motile bacteria detect specific chemical compounds and respond appropriately. The response is an altered swimming pattern resulting in net progress toward a more favorable chemical environment. This behavior, called *chemotaxis,* has been studied extensively in the closely related enteric bacteria, *Escherichia coli* and *Salmonella typhimurium.* Biochemical and genetic studies of those species have identified molecular components that mediate tactic behavior, including receptor proteins exposed on the cell surface, and delineated some of the mechanisms by which the components function. Studies of other bacterial species have revealed striking parallels with the tactic system of the enterics. It seems likely that chemotaxis in all bacterial species is accomplished by analogous sensory response systems that link receptors to flagella. In fact, homology has been detected among analogous protein components of the sensory systems from widely different bacteria. The current understanding of the chemotactic system in *E. coli* is summarized in the following paragraphs.

Chemotactic behavior

An *E. coli* cell swims in straight lines punctuated every few seconds by episodes of uncoordination, called *tumbles,* that result in new, randomly chosen directions of swimming. Thus the cell traces a three-dimensional random walk. Bacterial flagella are totally unlike eukaryotic flagella or cilia. The bacterial flagellum consists of a long helical protein filament, about the diameter of a single microtubule, that is turned like a propeller by a rotary motor embedded in the cell envelope. The motor is powered by proton motive force, not by adenosine triphosphate (ATP). A change in concentration of an active compound over either distance (spatial gradient) or time (temporal gradient) results in an alteration in tumble frequency. Favorable changes suppress tumbles; unfavorable ones induce them. A cell makes net progress along a spatial gradient by longer path lengths in favorable directions, tracing a biased random walk. Coordinated swimming in wild-type bacteria occurs when the left-handed helical filaments rotate counterclockwise, and tumbling occurs when they rotate clockwise. The sensory system influences cellular migration by controlling the direction of flagellar rotation and the probability of reversals.

The bacterial response to a gradient is transient. Addition of an attractant to a cell suspension results in an immediate suppression of all tumbles and addition of repellent induces continual tumbling. However, after a time ranging from seconds to several minutes, depending on the compound and the magnitude of the gradient, the cells resume their initial behavioral pattern of swimming and tumbling even though the attractant or repellent is still present. Thus, bacteria, like many sensory cells in higher organisms, adapt to the continued presence of a stimulus.

Excitation

Chemotactic behavior can be divided into two stages, excitation and adaptation. Excitation is fast, taking approximately 200 msec. It occurs when recognition of active compounds at receptor sites results in a shift in the balance between counterclockwise and clockwise rotation of the flagellar motor. The present understanding of excitation suggests that a change in occupancy of a ligand-binding site results in an alteration (an "excitatory conformation") in a transmembrane transducer protein, and that change in turn generates an "excitatory signal" that ultimately affects the flagellar motor.

The character of the excitatory alteration induced in a transducer upon ligand binding and the mechanism by which that change communicates information from transducer to motor are not known. There are six *che* gene products that are required for normal chemotactic behavior. Some *che* proteins may participate in the excitatory cascade. There are data arguing against direct physical interaction of transducers and motors in the plane of the membrane, observations indicating that the signal has a finite range not much greater than the length of a normal cell and indications that normal levels of cellular ATP are required for excitation. A good candidate for the excitatory signal would be a small molecule, perhaps phosphorylated, the intracellular concentration of which is controlled by the transducers.

Adaptation

Adaptation is the reestablishment of the initial balance between directions of flagellar rotation. It is relatively slow, requiring seconds to minutes. There is a convincing correlation between adaptation and methylation or demethylation of the transducers. It appears that the change in methylation level upon adaptation (each transducer has multiple methyl-accepting sites) effectively counteracts the excitatory conformational change induced by ligand binding, thus bringing the transducer back to a resting state. Adaptation to a stimulus mediated by a given transducer is correlated with an increase (positive stimuli) or a decrease (negative stimuli) in the level of methylation of that transducer. The donation of methyl groups from S-adenosylmethionine to specific glutamyl residues to form carboxyl methyl esters is catalyzed by the soluble *cheR* product and demethylation catalyzed by the soluble *cheB* product. In addition, four soluble proteins, the products of four other *che* genes, are required for normal adaptation as well as for maintenance of the swim/tumble balance in unstimulated cells.

Sensory response systems are designed to produce the appropriate response to relevant stimuli. This usually requires that the magnitude of a response be graded in proportion to the magnitude of the stimulus. Since many types of responses have an all-or-none character, the only way to grade a response is by controlling its duration. The larger the stimulus (i.e., the greater the number of occupied transducers), the longer the time required for the methylation system to neutralize the activated transducers and thus the longer the duration of the response. Any system with an all-or-none response would be expected to have an adaptation mechanism, and so it is not surprising that covalent modification of critical components is an emerging theme in the study of receptor systems.

Tactic sensitivities in enteric bacteria

Recognition sites for the two strongest attractants of *E. coli*, serine and aspartate, are located in the periplasmic domains of the transducer proteins Tsr and Tar, respectively. The attractant sugars galactose, ribose, and maltose are recognized by their respective periplasmic binding proteins, which are peripheral membrane proteins found in the space between the cytoplasmic membrane and the cell wall. Ligand-occupied maltose-binding protein interacts with the Tar transducer, generating excitation in a manner analogous to aspartate-Tar interaction, while the ribose- and galactose-binding proteins are linked to the Trg transducer. The enterics respond tactically to analogs that are recognized by the sugar or amino acid binding sites. In addition, the bacteria are attracted or repelled by a variety of compounds that appear to act by perturbing the cell in a way that affects some component(s) of the sensory motor system described here. For example, compounds that cause a decrease or an increase in intracellular pH are responded to as repellants or attractants, respectively. These compounds are not recognized by stereospecific binding sites, although the Tsr transducer must be present for the normal response to occur. Instead, it appears that protonation or deprotonation of a particular site on the intracellular domain of the Tsr protein affects the molecule in a way similar to excitatory changes upon decrease or increase of occupancy of the extracytoplasmic, serine-binding site. Adaptation occurs in the usual way by a change in the extent of methylation of the Tsr protein. Other compounds appear to act on the tactic system at steps after the transducers, and thus the methylation system is not involved in response to these chemicals. These observations may be relevant in considering complex chemosensory systems, like mammalian olfaction, in which a multitude of different compounds can be distinguished.

Transducers

The bacterial sensory system responds to temporal gradients. This indicates that a measure of concentration of an active compound at one moment (actually there is some integration over time) is compared to a record of the previous concentration. If the two differ, an appropriate signal is sent to the flagellar motors. We do not know how the comparison is accomplished, but the transducer molecules themselves have features which could provide the necessary functions. Occupancy of ligand-binding sites can measure current concentrations and the level of methylation (which changes more slowly than binding site occupancy) could provide a record of the just previous level of site occupancy. Imbalance between these two parameters would then induce the appropriate alteration of the "excitatory domain" of the transducer and generate

an "excitatory signal." Thus the bacterial sensory system exhibits properties of a rudimentary memory function. The transducer proteins appear to be intimately involved in comparison of the past and present.

Recent determination of the nucleotide sequences of transducer genes, *tsr, tar, trg,* and *tap,* revealed that they constitute a homologous gene family. Correlation of the deduced amino acid sequences of the transducers with biochemical information about sites of covalent modification have provided a simple model for the disposition of these homologous polypeptides across the cytoplasmic membrane. An N-terminal, ligand-recognition domain is located on the periplasmic side of the membrane, and a C-terminal domain containing sites of covalent modification and presumably the excitation domain is located on the cytoplasmic face. The two domains, which appear to fold as water-soluble structures, are connected by a single hydrophobic, membrane-spanning sequence about 40% of the way along the protein. The minimal connection between the two domains raises the issue of functional communication between them. The ligand-recognition domain might affect the cytoplasmic domain by movement transmitted along a single, membrane-spanning alpha-helix, or this first step in sensory transduction may involve interactions between individual transducer molecules in the plane of the membrane. It is interesting that the receptor for epidermal growth factor appears to be divided into two domains, one on each side of the plasma membrane, connected by a single stretch of hydrophobic amino acids, an organization analogous to that of the tactic transducers.

The four to six methyl-accepting glutamates found in the four transducers have been located in two regions of strong homology among all these proteins. One region, containing three or four methyl-accepting sites, is near the middle of the protein; the other, containing one or two sites, is near the C-terminal. The pattern of multiple methylation implies that the transducers adapt in a graded, quantized manner. A further complexity is that one or two of the methyl-accepting glutamates are created by deamidation of glutamines in a reaction catalyzed by the CheB demethylase. The occurrence of glutamines at sites destined to be methyl-accepting glutamates may function to create newly synthesized transducer proteins that are in an intermediate signaling state between the two extremes of modification (all negatively charged carboxyls versus all neutral methyl esters or amides) so that the overall sensory balance of an actively growing cell can be maintained.

The complexity and sophistication of the bacterial sensory system is striking. Elucidation of the molecular mechanisms underlying bacterial taxis is likely to have wide significance for the understanding of sensory receptor systems in general.

Further reading

Hazelbauer GL, Harayama S (1983): Sensory transduction in bacterial chemotaxis. *Int Rev Cytol* 81:33–70

Koshland DE Jr (1981): Biochemistry of sensing and adaptation in a simple bacterial system. *Ann Rev Biochem* 50:765–782

Macnab RM, Aizawa S-I (1984): Bacterial motility and the bacterial flagellar motor. *Ann Rev Biophys Bioeng* 13:51–83

Parkinson JS, Hazelbauer GL (1983): Bacterial chemotaxis: Molecular genetics of sensory transduction and chemotactic gene expression. In: *Gene Function in Prokaryotes,* Beckwith J, Davies J, Gallant JA, eds. Cold Spring Harbor Laboratory

Springer MS, Goy MF, Adler, J. (1979): Protein methylation in behavioral control mechanisms and in signal transduction. *Nature* 280:279–284

Cutaneous Sensory System

Lawrence Kruger

The integumentary surfaces are derived from ectoderm and consist of a variety of general and specialized regions of stratified epithelium with a superficial keratinized layer of varying thickness. The sensory function of the skin and its appendages (e.g., hairs and claws, which serve as mechanical levers) is achieved by means of a wide diversity of sense organs specialized for several distinct sensory modalities. Mechanical sensibility appears to be dominant, and sensations of touch, pressure, vibration, tickle, itch, with perhaps a contribution from position and movement (kinesthetic) sense, can be correlated with the properties of a variety of sensitive mechanoreceptors. Thermal sensations are conveyed by distinct populations sensitive to either warming or cooling. In many regions, the most numerous innervation consists of the thinnest fibers subserving sensory reports of pain or related noxious and aversive qualities.

There is a relationship between axon diameter (and therefore also conduction velocity) and each class of sense organ, but this is not an absolute scheme because of shifts in fiber spectrum, e.g., larger axon diameters and fewer unmyelinated axons rostral in the neuraxis, fewer unmyelinated axons distally on the limbs, and zones of regional specialization (e.g., the cornea and dental pulp) possessing a restricted range of thin axon diameters.

Sensitive mechanoreceptors appear to display the richest variety of morphological and functional patterns, but all can be defined as velocity detectors with a specific range of sensitivity. One class of velocity-sensitive sense organ also responds to sustained, constant displacement, with impulse discharge number and frequency a function of stimulus amplitude and force; this class is called slowly adapting, although these sense organs can respond to high-frequency repetitive displacement. The slowly adapting receptors are believed responsible for sensations of tactile magnitude or pressure, and because of their high velocity sensitivity, they probably also contribute to transient and vibratory sensibility. Two main types are recognized: (1) Merkel cell neurite complexes (type I) in the basal epidermis, and (2) Ruffini endings (type II) embedded in connective tissue, responsive to lateral stretch, and exhibiting a distinctive impulse discharge pattern of regular intervals. Both types are innervated by large myelinated axons. Type I receptors are often associated with epidermal protuberances (domes) or large specialized hairs, but their sensory role remains obscure.

The dermis contains a variety of sense organs responsive only during the velocity phase of displacement. Those associated with the larger guard hairs relay via faster conducting axons and a different spinal pathway than those associated with the finer down hairs. Further subdivision has been suggested on the basis of the range of velocity sensitivity. The velocity detectors of glabrous skin are similar to those activated by hairs, and it can be inferred that many of these endings display a laminar morphology called corpuscular endings. The best-studied and most elaborate corpuscular sensory structures, called Pacinian bodies or endings, are exquisitely sensitive to repetitive tactile stimuli and may act as detectors of velocity change (i.e., acceleration) contributing to vibratory sensibility.

Thermal sensibility is represented by separate populations of cold and warmth sense organs (or receptors) innervated by the thinnest myelinated and most by unmyelinated (C) axons. Sense organs for detection of cooling are distributed in distinctly punctate fashion, cold spots, and are believed to be located in the basal epidermis. The structure and properties of the far less numerous warmth detectors remain obscure.

The unmyelinated (C) axons constitute the largest group of most sensory nerves and are dominated by a variety of sense organs called nociceptors believed to underlie pain sensibility. One class of nociceptor is supplied by thinly myelinated axons ending in pain spots in the basal epidermis and inferred to convey fast, pricking pain. The larger class, consisting of C fiber nociceptors, probably subserves slow, burning, prolonged, and chronic pain and constitutes a range of morphological and functional varieties. Many of these sensory axons appear to be involved in efferent control (axon reflex) of the vasodilation and plasma extravasation, believed to be largely neuropeptide mediated, subserving the process of neurogenic inflammation and the increased sensitivity that slowly develops after local tissue damage—a phenomenon unique to nociceptors called sensitization. Some C nociceptor axons are believed to enter the ventral roots, but the diversity of peptidergic and other chemical specialization unique to this class of sense organs have not been categorized in terms of their pathways.

The central pathways of the varieties of cutaneous sense organs become segregated at the dorsal root entry zone of the spinal cord. Nociceptor and thermoreceptor inputs via the lateral dorsal root division synapse widely at the entry level and project predominantly via the contralateral anterolateral funiculus (and trigeminal homolog for the head). The sensitive mechanoreceptor projection is predominantly via the medial division of dorsal roots entering the ipsilateral dorsal column without synapse, but there is a substantial polysynaptic tactile contribution to both lemniscal and anterolateral systems.

The dorsal column axons project principally upon the ipsilateral cuneate and gracile nuclei, the cells of which emit axons that decussate and end primarily in the contralateral thalamic ventrobasal complex, with smaller contributions to other zones (e.g., superior colliculus). The anterolateral system, by contrast, consists primarily of second-order axons derived from the contralateral spinal grey. This system contributes a large medullary and mesencephalic projection and a smaller thalamic (spinothalamic) projection, largely overlapping the termination zone of the lemniscal system and a distinctive zone within several nuclei of the intralaminar wing (e.g., submedius and centralis lateralis) and a small sector of the posterior group. The organizational pattern of each subcortical somatosensory zone appears to be dominated by a somatotopic arrangement, although there is some evidence for modality segregation.

The thalamic projection to the cerebral cortex is primarily to the postcentral gyrus in primates and secondarily to the invariably contiguous motor area and to caudal association areas. The pattern of sensory cortex projection consists of multiple somatotopic representations, and there is some evidence for zones or vertical modules segregating specific sense organ projections. These modules, sometimes called columns of skin vs. deep afferent projections, appear to display different varieties of integrative properties and to be dominated by a specific sensitive mechanoreceptor projection, but the function of vertical modules and their form and dimensions remain undetermined. The projection of modalities represented by the more numerous thin-fiber cutaneous sense organs (e.g., thermal and nociceptive) remains obscure and is largely known only in terms of convergent patterns.

Further reading

Sinclair D (1981): *Mechanisms of Cutaneous Sensation*. Oxford: Oxford University Press

Willis WD, Coggeshall RE (1978): *Sensory Mechanisms of the Spinal Cord*. New York: Plenum Press

Echolocation

Nobuo Suga

The order *Chiroptera* (bats), comprising one-fifth of mammalian species, consists of two suborders: Megachiroptera (154 species) and Microchiroptera (about 800 species). All microchiropterans thus far studied echolocate, but only one megachiropteran does. Morphology and ecology of bats are so diverse that *echolocation* behavior and orientation sounds (pulses) are quite different among different species of bats. Studies in neuroscience of echolocation have been mainly performed with three different species of microchiropterans: *Myotis lucifugus* (little brown bat), *Pteronotus parnellii* (mustached bat), and *Rhinolophus ferrumequinum* (horseshoe bat). Among them, the functional organization of the auditory system is different, reflecting the unique acoustic properties of the bat's own orientation sounds, although they share the same anatomical structures and neural mechanisms for echolocation to a great extent. The parallel hierarchical processing of biosonar information was first and best explored in *Pteronotus*, which emits complex orientation sounds.

Properties of biosonar signals

For prey (e.g., insect) capture and orientation, microchiropterans emit constant frequency (CF) and/or frequency-modulated (FM) sounds. The orientation sound of *Pteronotus* always consists of a long CF component followed by a short FM component (Fig. 1A). Since each orientation sound contains four harmonics (H_{1-4}), there are potentially eight major components (CF_{1-4}, FM_{1-4} in Fig. 1, C1). The second harmonic (H_2) is always predominant, with CF_2 at about 61 kHz and FM_2 that sweeps from 61 kHz to about 49 kHz (Fig. 1A). The third harmonic (H_3) is 6–12 dB weaker than H_2, while the first and fourth harmonics (H_1 and H_4) are 18–36 dB and 12–24 dB weaker than H_2, respectively (Fig. 1B). Such a species-specific orientation sound can be elicited by electrical stimulation of the periaqueductal gray or reticular formation of the midbrain.

Echoes that elicit behavioral responses in *Pteronotus* always overlap with the emitted sound, so that biosonar information must be extracted from a complex sound containing up to 16 components. Different components or parameters of echoes carry different types of target information (Table 1). The long CF component is an ideal signal for target detection and measurement of target velocity, while the short FM component is more appropriate for ranging, localization, and characterization of a target. *Pteronotus* and *Rhinolophus* use the long CF signal to obtain velocity information, in particular, insect wing-beating velocity. The acquisition of velocity information is optimized by a unique acoustic behavior called Doppler shift compensation.

The auditory system

The gross structure of the auditory system of *Pteronotus* is basically the same as that of other mammals, but it shows a unique functional organization reflecting the properties of the orientation sound and echolocation behavior. One of the most striking features is the sharp tuning of the auditory periphery for fine frequency analysis of the CF_2 component. Since the cochlea has resonators sharply tuned to 61 kHz (the frequency of the CF_2), the frequency-tuning curves of auditory nerve fibers tuned to 61 kHz have a quality factor of 210 and a slope of 1500–1900 dB/octave. Because of this extremely sharp tuning, they can code a 6-Hz frequency shift from 61 kHz. That is, they can easily code the small frequency modulation that would be evoked by the wings of flying insects. The CF_2 processing channel is disproportionately large, from the cochlea through the auditory cortex.

The other striking feature of the auditory system of *Pteronotus* is the parallel hierarchical organization to extract certain types of biosonar information by combination-specific neurons. All the CF and FM components of the orientation sound differ in frequency (Fig. 1, C1), so that they are separately analyzed in the cochlea and are sent in parallel into the brain by different auditory nerve fibers. In the CF_{1-3} processing channels, many neurons at higher levels show sharp "level-tolerant" frequency tuning for further analysis of the CF signals. In a level-tolerant tuning curve, the bandwidth of the curve stays very narrow regardless of stimulus levels. The level-tolerant tuning results from lateral inhibition, which works most dramatically for the CF_2 processing channel. The parts of the CF_{1-3} processing channels are integrated in the medial geniculate body for extraction of velocity information from orientation sound-echo pairs. As a result, CF/CF combination-sensitive neurons are produced. These neurons project to the CF/CF area in the auditory cortex (Fig. 2B, c). In the FM_{1-4} processing channels, some neurons at higher levels respond selectively to FM sounds because of neural circuits incorporating disinhibition. The parts of the FM_{1-4} processing channels are integrated in the medial geniculate body for extraction of range information from orientation sound-echo pairs. As a result, FM-FM combination-sensitive neurons are produced. These neurons project to the FM-FM area in the auditory cortex (Fig. 2B, b). As a consequence of such parallel hierarchical processing of complex sound, there are many functional divisions in the auditory cortex (Fig. 2B).

In the primary auditory cortex of *Pteronotus*, high-frequency-sensitive neurons are located anteriorly and low-frequency-sensitive neurons are located posteriorly, as in the cat (Fig. 2C). But the auditory cortex of the bat shows at least two unique aspects: (1) About 30% of this cortex is devoted to the representation of 61–63-kHz sound, although the total range of frequency representation ranges from 7 to 120 kHz. The 61–63-kHz tuned area is called the DSCF area (Fig. 2B, a). (2) Another 30% of the cortex is devoted to extract target velocity and range information from orientation sound-echo

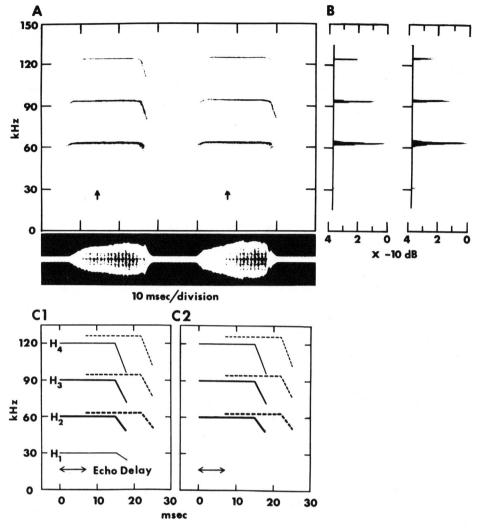

Figure 1. Sonagram of orientation sounds (A), the amplitude spectra of the CF components in these sounds (B), and schematized sonagrams of the orientation sound and echo pairs (C). A. Sonagram (upper figure) and wave form (lower figure) of two orientation sounds emitted by the mustached bat in flight. Faint harmonics are enhanced for clarity in the figure. B. The amplitude spectra of the CF components of the orientation sounds (arrows in A). Note that the first harmonic is the weakest among the four harmonics (about 36 dB weaker than the second harmonic), but it is one of the essential signal elements for facilitation of responses of CF/CF and FM-FM combination-sensitive neurons. C. Schematized sonagram of two pairs of orientation sounds (solid lines) and echoes (dashed lines). When the bat emits an orientation sound, the first harmonic stimulates the animal's own ears mainly by bone conduction. Then, CF/CF and FM-FM neurons are conditioned to respond to an echo with a specific amount of Doppler shift or echo delay (C1). If the first harmonic is not radiated at all, no combination of orientation sounds or echoes produced by conspecifics would evoke a facilitative response in CF/CF and FM-FM neurons of the bat that merely hears such sounds (C2). As a result, jamming of echolocation by conspecifics would be significantly reduced. H_{1-4} are the first through fourth harmonics. The CF and FM components in the H_{1-4} are called CF_{1-4} and FM_{1-4}. From Suga (1984).

Table 1. Different Acoustic Parameters of an Echo Relative to the Orientation Sound at the Ear Carry Different Types of Target Information

Echo	Target
Doppler shift	velocity
DC component	relative velocity
AC component	flutter
amplitude	subtended angle
delay	range
amplitude + delay	size
amplitude spectrum	fine characteristics
binaural cues	azimuth
"pinna-tragus" cue	elevation

pairs by comparing two signal elements. This combination-sensitive area consists of two major divisions: CF/CF and FM-FM (Fig. 2). The DSCF, CF/CF, and FM-FM areas show unique functional organization for representation of different types of biosonar information, as summarized in Fig. 2C.

Neurons in the DSCF area are tuned to particular frequencies and amplitudes and form the frequency vs. amplitude coordinates. Their frequency tuning is extremely sharp and level-tolerant, being related to the need for level-tolerant frequency (velocity) discrimination. Both level-tolerant frequency tuning and amplitude selectivity result from lateral inhibition. The frequency axis of the DSCF area matches the CF_2 frequencies of the bat's own orientation sound and Doppler-shifted echoes.

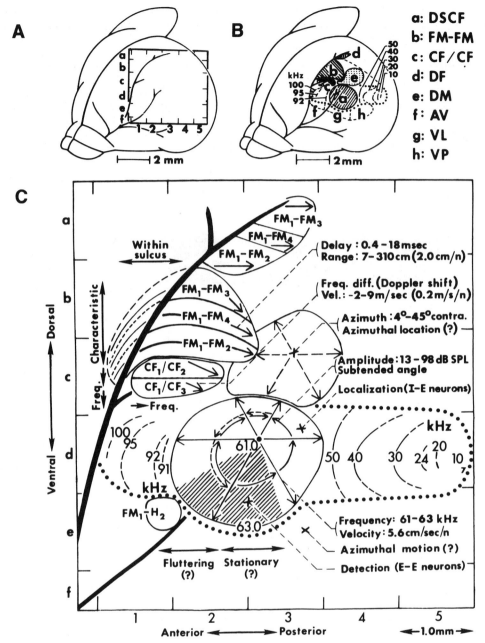

Figure 2. The functional organization of the auditory cortex of the mustached bat. A. The dorsolateral view of the left cerebral hemisphere and the branches of the median cerebral artery. The long branch is on the sulcus. The area within the rectangle consists of several divisions, as shown in B. The functional organization of certain divisions have been electrophysiologically explored, as shown in C. B. The several divisions of the left auditory cortex (a-f). The DSCF, FM-FM, CF/CF, and DF areas are each specialized for the systematic representation of biosonar information. The functional organization of these areas are graphically summarized in C. C. A graphic summary of the functional organization of the auditory cortex. The tonotopic representation of AI and the functional organization of the DSCF, FM-FM, and CF/CF areas are indicated by lines and arrows. The DSCF area has axes representing target-velocity information (echo frequency: 61–63 kHz) or subtended target-angle information (echo amplitude: 13–98 dB SPL) and is divided into two subdivisions suited for either target detection (shaded) or target localization (unshaded). These subdivisions are occupied mainly by E-E or I-E neurons. The anterior and posterior halves of the DSCF area are hypothesized to be adapted for processing echoes from either fluttering or stationary targets.

The FM-FM area consists of three major types of FM-FM neurons (FM$_1$-FM$_2$, FM$_1$-FM$_3$, and FM$_1$-FM$_4$), which form separate clusters. Each cluster has an axis representing a target range from 7 to 310 cm (echo delay: 0.4–18 msec). The dorsoventral axis of the FM-FM area probably represents fine target characteristics. The CF/CF area consists of two major types of CF/CF neurons (CF$_1$/CF$_2$ and CF$_1$/CF$_3$), which are also found in separate clusters. Each cluster has two frequency axes and represents a target velocity from −2 to +9 m/sec (echo Doppler shift: −0.7 to +3.2 kHz for CF$_2$ and −1.1 to +4.8 kHz for CF$_3$). The DF area and a part of the VA area receive a projection from the FM-FM area. The DF area consists of three clusters of FM-FM neurons and has an axis representing a target range from 7 to 140 cm, whereas the VA area contains only a single cluster of FM$_1$-H$_2$ neurons. The DM area has an axis representing the azimuth location of the target on the contralateral side in front of the animal. This azimuth representation is incorporated with tonotopic representation. In the VP area, azimuth motion-sensitive neurons have been found. The functional organization of the VA and VP areas remains to be further studied. From Suga (1984).

The amplitude representation by the location of excited neurons is called amplitopic representation.

Neurons in the CF/CF area respond poorly to orientation sounds (pulses) and echoes when presented separately, but show the remarkable facilitative response to a pulse-echo pair. Because of the facilitation, CF/CF neurons dramatically decrease the threshold of response to the echo. They are tuned to particular amounts of Doppler shifts. The essential elements in the pair is the CF_1 of the pulse and the CF_2 or CF_3 of the echo. For maximum facilitation, a precise frequency relationship between the two CF signals is required regardless of stimulus levels, because the majority of CF/CF neurons show level-tolerant frequency tuning for facilitation. The frequency axes of the CF/CF area for facilitation match the CF frequencies of the bat's own orientation sound and Doppler-shifted echoes. In the CF/CF area, the discharges of 35% of neurons can phase-lock to periodic frequency modulations of the echo CF component that would be evoked by the wings of a flying insect. However, the remaining 65% show no phase-locked discharges to such a stimulus. The first group of neurons is extremely interesting because their discharges are phase-locked to the periodic frequency modulations only when the pulse is paired with a Doppler-shifted echo.

In the FM-FM area, neurons respond poorly or not at all to either a pulse or echo alone, but show the strong facilitative response to a pulse-echo pair with a particular echo delay. Because of the facilitation, the threshold of response to the echo dramatically decreases. The essential elements in the pulse-echo pair are the FM_1 of the pulse and the FM_{2-4} of the echo. FM-FM neurons are tuned to particular echo delays (best delays) and are systematically arranged according to the values of their best delays. Since echo delay corresponds to target range, the echo delay axis is the range axis. The systematic representation of target range by the location of excited neurons is called odotopic representation. About 25% of FM-FM neurons are tuned to echo delays shorter than 4 msec (69-cm target range) and respond vigorously to each pulse-echo pair delivered even at a rate of 100/sec. Therefore, the FM-FM area of the auditory cortex is involved in information processing even during the terminal phase of echolocation.

FM-FM neurons do not show facilitation of responses to unnaturally intense echoes. They are tuned not only to particular echo delays but also to particular echo amplitudes. Therefore, they would respond best to targets of particular cross-sectional areas at particular distances. FM_1-FM_4 neurons are theoretically better suited for fine characterization of small targets than FM_1-FM_2 neurons because FM_4 is much shorter than FM_2 in wavelength and has broader bandwidth. Thus the distribution of neural activity perpendicular to the range axis on the cortical plane represents the amplitude spectra of all FM components of an echo, which are directly related to target characteristics such as shape and size.

Protection of the neural representation of biosonar information from jamming by orientation sounds of conspecifics

The mustached bat lives in large colonies. Consequently, one important problem faced by individual bats is how to protect their echolocation system from the jamming effect of biosonar sounds produced by conspecifics. Seven possible mechanisms could work together to reduce jamming: (1) the sharp directionality of the orientation sound, (2) the sharp directional sensitivity of the ear and binaural hearing, (3) the signature of orientation sounds and the personalized auditory system, (4) an auditory time gate, (5) heteroharmonic sensitivity, (6) variation in emitted signals and the sequential processing of echoes, and (7) efferent copy originating from the vocal system. (4) and (5) are related to the response properties of FM-FM and CF/CF neurons (Fig. 1, C2).

Further reading

Suga N (1984): The extent to which biosonar information is represented by the auditory cortex of the mustached bat. In: *Dynamic Aspects of Neocortical Function*. Edelman GM, Cowan WM, Gall WE, eds. New York: John Wiley & Sons, pp 315–373

Busnel RG, Fish JF, eds. *Animal Sonar Systems* New York: Plenum (1980)

Suga N, O'Neill WE, Kujirai K, Manabe T (1983): Specialization of "combination-sensitive" neurons for processing of complex biosonar signals in the auditory cortex of the mustached bat. *J Neurophysiol* 49:573–1626

Electric Organs, Fishes

Michael V.L. Bennett

Electric organs, found in six groups of fishes (Fig. 1), are structures specialized to generate electric fields in the animals' external environment. In some the voltages are large enough to stun prey or repel predators. These, the strongly electric fishes, include the electric eel (*Electrophorus electricus*) from South America, the electric catfish (*Malapterurus electricus*) from Africa, the family of electric rays, the Torpedinidae, which are cosmopolitan and marine, and possibly the stargazers (*Astroscopus* sp.) of the Western Atlantic.

The weakly electric fishes emit lower voltages that have no direct offensive or defensive role. These pulses serve as the energy source in an active electrolocation system that monitors the electrical impedance of environment. These weak signals can also serve in communication within and between species. There are three groups of weakly electric fishes, the South American knife fishes, the Gymnotiformes, which include the electric eel, the African Mormyriformes and the cosmopolitan marine skates, the Rajidae. Each group contains many species. The electric eel and some torpedinids have both weakly and strongly electric organs.

The strongly electric fishes produce remarkably powerful pulses. A large electric eel generates in excess of 500 V. A large Torpedo generates a smaller voltage, about 50 V in air, but the current is larger and the peak pulse power in each case can exceed 1 kW. The electric organs in these species makeup a substantial fraction of the body mass.

Weakly electric fishes generate signals from a few tenths of a volt to a few volts, but the signals are still large compared to what is generated by nonelectric fishes. Some species continually emit pulses at frequencies that are remarkably high, up to at least 1400/sec.

Electrocyte operation

The generating elements of the electric organs are specialized cells termed *electrocytes* (or electroplaques or -plax in those species where the cells are flattened). Their basic membrane properties are similar to those of other excitable cells, in particular the muscle fibers from which they are evolved. Vestiges of muscle filamentous structure remain in some species but no electrocytes are contractile. In one gymnotiform group the electrocytes are modified nerve fibers, the myogenic electric organ having been lost.

The basic operation of an electrocyte involves apposed membranes with different properties. In the simplest case each membrane at rest has the same resting potential, determined by potassium permeability and concentration gradient. Activation of one membrane by neural inputs from the central nervous system greatly increases its permeability to sodium ions. A large influx of Na ions ensues, which makes the inside more positive and causes efflux of K ions through the opposite face. All electrocytes operate in this manner; apposed membranes have different permeabilities, and during activity ions flow down their concentration gradients to generate electric currents. Metabolic energy is expended subsequently to restore the ionic inequalities. The currents generated by single cells summate both in series and in parallel. The large outputs of electric organs are not due to qualitative differences in their excitable membranes. Rather, the membranes have higher conductances due to the presence of more channels, the apposed membranes have different properties to maximize external current flow, many cells act in series and parallel, and finally the organs can be arranged within the body to increase current flow in the environment.

Electrocytes of different species have different arrangements

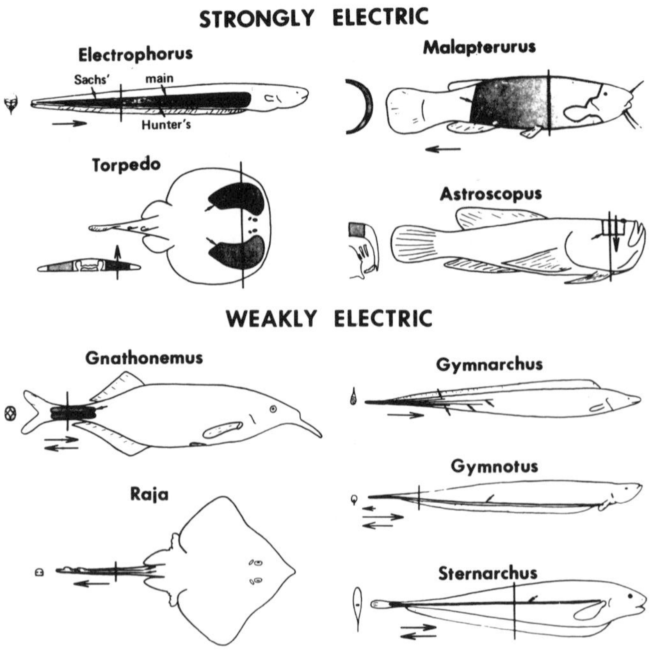

STRONGLY ELECTRIC

WEAKLY ELECTRIC

Figure 1. Representative electric fishes. All are shown from the side except *Torpedo* and *Raja,* which are shown from the top. Electric organs are stippled or solid and indicated by small arrows. Cross-sectional views through the organs are also shown at the levels indicated by the solid lines. The large arrows indicate the direction of active current flow through the organs; where there is more than one arrow, they indicate successive phases of activity, and the relative lengths are proportional to the relative amplitudes of the phases. *Gnathonemus* is a mormyrid. From Bennett MVL (1968): Neural control of electric organs. In: *The Central Nervous System and Fish Behavior,* Ingle D, ed. Chicago: University of Chicago, pp 147–169.

of excitable membranes, which can produce electric organ discharges of different pulse shapes (Fig. 2). In strongly electric fishes the electrocytes are always flattened. In the marine forms, the torpedos and stargazers and some skates, the innervated face of the cells generates only excitatory postsynaptic potentials (EPSPs) (or end plate potentials). The innervation is profuse and much of the surface is subsynaptic membrane. Torpedo electric organ is a prime source of acetylcholine (ACh) receptor protein. As at neuromuscular junctions, the permeability increase is to Na+ and K+ and the reversal potential of the EPSP is near OmV. The uninnervated face of these cells is inexcitable and has at rest a much higher conductance than

the innervated face. However, when the innervated face becomes active, its conductance greatly increases. Thus, there is a form of impedance ance matching that maximizes efficiency at the organs.

In the electric eel and several weakly electric relatives, the innervated face generates an Na-dependent action potential in addition to EPSPs. As in the marine species the uninnervated face is of much higher conductance than the innervated face when inactive, but the conductances become more similar during activity. Most of the electrocyte response is due to the Na mechanism.

In the eel the depolarization that increases Na permeability also decreases permeability of anomalously rectifying K channels. This change reduces eddy currents in the innervated face. In the eel the action potential in the innervated face is terminated by Na inactivation, which is simply closure of the Na channels in spite of maintained depolarization. In other species there is also voltage-dependent activation of K channels which speeds restoration of the membrane potential to its resting value.

In many freshwater fishes, which have Na-dependent action potentials, the innervation occurs on a stalk or stalks of varying degrees of complexity coming off one face (functionally the innervated one). Impulses are initiated in the stalks and propagate to excite the main part of the electrocytes. The functional significance is unclear, but presumably stalks allow substitution of Na channels for ACh-activated channels.

In many weakly electric fishes in mormyriform and gymnotiform groups, both faces of the electrocytes generate impulses. First the innervated face generates its impulse and current flows in one direction. This activity excites the opposite, uninnervated face and current flows in the other direction. Thus, a biphasic external potential results. In some species a third phase of discharge can be generated by activity of the stalk systems. In the electric catfish the impulse of the uninnervated face is initiated with little delay and outlasts the brief impulse of the innervated face. The major current flow is due to the uninnervated face and the response is almost monophasic. The impulse in the innervated face presumably serves to increase its K conductance and thereby allow a larger current flow during the impulse of the uninnervated face. A similar mechanism is found in some skates in which depolarization of the uninnervated face increases its K conductance and thus increases the current flow caused by the EPSPs generated in the innervated face.

In electrocytes of a few weakly electric fishes the uninnervated face acts as a series capacity. Current flows outward during the rising phase of the impulse in the innervated face and inward during the falling phase. Repetitive activity of the innervated face thus produces an alternating current externally.

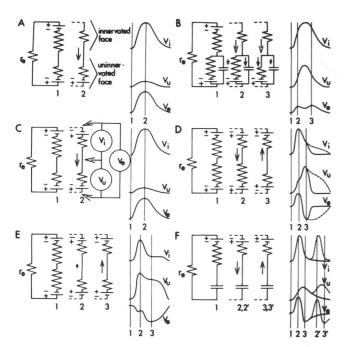

Figure 2. Activity of different kinds of electrocytes and their equivalent circuits. The upper part of each circuit, labeled only in A, represents the innervated or stalk face; the lower part represents the uninnervated or nonstalk face. The resistance of the extracellular current pathway is represented by r_e. The potentials that are recorded differentially across the two faces (V_i and V_u) and across the entire cell (V_e) are drawn to the right of each circuit (intracellular positivity and positivity outside the uninnervated face shown upwards; $V_i - V_u = V_e$). Placement of electrodes for recording these potentials is indicated in C. The successive changes in the membrane properties are shown by the numbered branches of the equivalent circuits, and their times of occurrence are indicated on the potentials. A lower membrane resistance is indicated by fewer zigzags in the symbol. Return to resting condition is omitted. A. Electrocytes of strongly electric marine fish and some skates. The innervated face generates only a PSP. B. Electrocytes of other rajids. The innervated face generates a PSP and the uninnervated face exhibits delayed rectification. C. Electrocytes of the electric eel and a few other gymnotids. The innervated face generates a Na-dependent impulse. D. Electrocytes of mormyrids and some gymnotids. Both faces generate a spike and V_e is diphasic. In some the spike across the uninnervated face is longer lasting and the second phase of V_e predominates. These potentials are shown by dotted lines. E. Electrocytes of the electric catfish. The stalk face is of higher threshold and generates a smaller spike (indicated by a smaller battery symbol) and the external potential is largely negative on the nonstalk side distant from the stalk. F. Electrocytes of *Gymnarchus*. The uninnervated face acts as a series capacity and the external response has no net current flow. The summation of a second response ($2_i,3_i$) on the first is shown by the dotted lines. Probably electrocytes of several gymnotids operate similarly. From Bennett MVL (1970): *Annu Rev Physiol* 32:471–527.

Discharge patterns

Strongly electric fishes emit monophasic pulses that arise from synchronous activity of their electrocytes (Fig. 3). Monophasic pulses should be more effective stimulators. Not surprisingly in view of the large power outputs, the organs are discharged only intermittently. Freshwater weakly electric fishes, the mormyriforms and gymnotiforms, discharge their organs continually at frequencies from a few per second to over 1400 per second. In most species the discharge is due to synchronous activity of the electrocytes and thus is biphasic or even triphasic. Both groups can be divided into two types according to pattern of discharge, the pulse species in which the pulses are brief compared to the interval between them and the wave

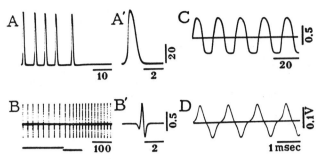

Figure 3. Patterns of electric organ discharge. A,A′. An electric catfish 7 cm long recorded from in a small volume of water with head negativity upward. Mechanical stimulation evoked a train of five pulses which attained a maximum frequency of 190/sec. A′. Single pulses could also be evoked (faster sweep speed). B-D. Weakly electric gymnotids recorded immersed in water, head positivity upward. B. A variable frequency gymnotid, *Gymnotus;* pulses were emitted at a basal frequency of 35/sec. Tapping the side of the fish at the time indicated by the downward step in the lower trace caused an acceleration up to about 65/sec. The acceleration persisted beyond the end of the sweep. The small changes in amplitude are due to movement of the fish with respect to the recording electrodes. B. Faster sweep showing the triphasic pulse shape. C. A constant, low frequency gymnotid, *Sternopygus*. The pulse frequency is about 55/sec. The horizontal line indicates the zero potential level. D. High-frequency wave type gymnotid, *Sternarchus*. The frequency is about 800/sec. The horizontal line indicates the zero potential level. Calibrations in volts and milliseconds. From Bennett MVL (1970): *Annu Rev Physiol* 32:471–527.

species in which the pulse and interpulse duration are comparable. Pulse species tend to increase their discharge frequencies in response to a variety of stimuli; wave types are more constant although they will change in response to social signals or to move away from the frequency of another fish (jamming avoidance response).

Responses that can be evoked by handling skates, the marine weakly electric fishes, involve asynchronous and repetitive activity of the electrocytes; their discharges under normal conditions are poorly known.

Electric organ discharges are generally controlled by a central command nucleus whose neurons discharge synchronously to cause each electric organ pulse. The cells are electrotonically coupled, which is responsible for the synchronization. The command neurons integrate various sensory and central inputs and are probably spontaneously active in many species. The command volley is in most fishes relayed by lower level nuclei, which in higher frequency species are electrotonically coupled also, presumably to increase synchronization. In mormyrids the command nucleus sends a signal (efference copy) to sensory nuclei to modulate their responses to afferent activity evoked by organ discharge.

Evolution

Electrosensory systems have importance both in understanding electric organ discharges and in accounting for the evolution of strongly electric organs. The significance of discharges with little direct current component is that it allows the fish to have two sets of electroreceptors, one that is sensitive to low frequencies of environmental origin and one that is sensitive to the high frequencies of the electric organ discharges and used in electrolocation and communication. Diversity of discharge types also results at least in part from evolution of mechanisms of species recognition in a crowded environment. Darwin supposed that weakly electric organs were intermediates in the development of strongly electric organs, but he could see no value in the weakly electric organs. Later, Hans Lissmann, who initially demonstrated active electrolocation, proposed this function as selecting for the weakly electric organs which then attained sufficient voltage to lead to the strongly electric organs.

Electrocution

An often asked question is whether strongly electric fishes electrocute themselves. They don't and a major preventive factor is the presence of large amounts of connective tissue and fat around sensitive structures such as nerves. Whether their brains are more resistant to electroconvulsive shock is unknown.

Value of electric organs to physiology

Historically electric organs demonstrated that Galvani was essentially correct about animal electricity. In modern times also they have been the source of a large body of biophysical and comparative data. Because the evolutionary pressures led to a large volume of tissue with relatively pure excitable membranes, electric organs have been important starting materials in the biochemical characterization of macromolecules such as the acetylcholine receptor and the sodium channel.

Further reading

Bennett MVL (1971): Electric organs. In *Fish Physiology* Hoar WS, Randall DJ, eds. 5:347–391.
Bennett MVL (1971): Electroreceptors. In *Fish Physiology*, Hoar WS, Randall DJ, eds. 5:493–574.
Bullock TH, Heiligenberg (1985): *Electroreception* New York: Wiley and Sons.

Electroreceptors and Electrosensing

Theodore H. Bullock

Sense organs with a specialized sensitivity to very feeble electric fields in the water are found in most nonteleost fish (except hagfish and holosteans) and in four orders of teleosts, widely scattered in the skin and innervated by special branches of the lateral line nerves. They can be designated *electroreceptors* based on evidence that they respond to naturally occurring electric currents of obvious biological significance to the organism and that they are the necessary mediators of normal behavioral responses to electrical events. Some electroreceptors are quite sensitive, phasically, to temperature change and poorly sensitive to mechanical pressure on the skin, but to what degree these sensibilities are due to electric consequences is not known.

Two general classes of electrosense organs are the ampullary and the tuberous organs; each has several varieties. Besides gross and fine anatomical differences, the ampullary receptors are sensitive to low-frequency (0.2 to 15 Hz) sinusoidal fields, the tuberous to higher frequency fields. In different species the working range of tuberous organs varies between 50–200 Hz and 5000–18,000 Hz.

Ampullary organs have apparently been invented at least twice. One type is found and is presumably homologous through the petromyzoniforms, elasmobranchs, holocephalans, polypteriforms, chondrosteans, dipneustans, crossopterygians, and some apodan and urodele amphibians; these are sometimes called Lorenzinian ampullae. They are excited by a cathode near the external opening, i.e., by depolarization of the apical membrane of the receptor (modified hair) cell. Another type, called teleost ampullae, is found in apparently all species of siluriforms, gymnotiforms, mormyriforms, and one subfamily (xenomystines) of osteoglossiforms. These are excited by an anode close to the apical end of the receptor cell, i.e., by depolarization of the basal membrane. Both types presumably function primarily to detect weak electric gradients due to unspecialized (not electric organ) animate as well as inanimate sources.

Tuberous receptors are found only in electric teleosts of the orders Gymnotiformes (tropical South and Central America) and Mormyriformes (tropical Africa) all of which have some ongoing rate of electric organ discharge (EOD) whose waveform is characteristic of the species and often the individual fish. The power spectrum maximum is matched in each individual to the receptors' best frequency. Tuberous organs in individual fish may have narrow bands, with a sharp best frequency; most units in a given fish have virtually the same best frequency. The tuberous organs of gymnotiforms and mormyriforms are different anatomically and presumed to be the result of parallel evolution; this applies also to the whole system of central electrosensory nuclei and pathways. Two distinct kinds of tuberous receptors are generally present in each species; the distinctions are not the same in the different families, but may include sensitivity, dynamic range, best stimulus parameter, and regularity. One basis for differences is the way each subclass encodes gradation of stimulus intensity into afferent fiber spikes. The variety of coding types emphasizes that the nervous system uses several nerve impulse codes, not only a single code.

Information is available on the role of sex steroids in the tuning of tuberous receptors in gymnotiforms with a sex-correlated difference in EOD rate. Data is available on the ontogeny including the increase throughout life in the total number of organs and in the number per axon.

Electrosensing by behavioral endpoints is best known in four contexts. (1) Detection of homogeneous fields in sea water by sharks and rays is the most sensitive; thresholds can be as low as 0.005 μV/cm. Freshwater rays and catfish are significantly less sensitive in terms of voltage gradient. (2) Detection of magnetic fields by the electric current induced through relative motion of the elasmobranch as part of a flowing ocean stream, or as a swimming object, extends down to intensities weaker than earth strength, even at the slow motion of ocean streams. (3) Active electrolocation in electric fish by the detection and discrimination of objects that distort the fish's own EOD field is useful within approximately a body length, over a broad optimum of water conductivity. (4) Electrocommunication by detection of the EOD of other fish, especially of the same species, is useful in social signaling over distances of several body lengths and is specialized for the discrimination of small differences in the amplitude or shape of brief pulses (0.3 msec) in the species with brief EODs below 100/second ("pulse species") or amplitude and phase modulations of the fish's own field in the species with a quasisinusoidal EOD ("wave species"). In the last group behavioral response shows a sensitivity to phase shift down to <0.5 μsec. There may be other roles of electroreception in the ecology of the widely disparate taxa, with widely diverse environments and habits of life—saltwater and fresh, clear as well as murky water, diurnal habits as well as nocturnal, predators, herbivores, and scavengers.

A great deal is known about the anatomy and physiology of the central nuclei and pathways, especially in gymnotiforms and mormyriforms, from the medulla up through the midbrain and cerebellum. More than 30 cell types are known to be involved in processing this input. One form of social response, the jamming avoidance response, has been studied in particular and neurons are known in detail through at least 14 orders from receptors to effectors, accounting quantitatively for the observed behavior.

Further reading

Bullock TH (1982): Electroreception. *Annu Rev Neurosci* 5:121–170

Heiligenberg W, Bastian J (1983): The electric sense of weakly electric fish. *Annu Rev Physiol* 46:561–583

The Genital Sensory System

James D. Rose

The genital sensory system has a unique constellation of properties, including sensory discriminative operations typical of somatic sensation, the affective and motivational characteristics of genital sensation, and various behavioral and endocrine control functions vital to reproduction. Genital sensory function is also unique in that sexual dimorphism is most extreme in the genital organs, thus providing the brain with a notably different view of the body periphery in the two sexes.

Peripheral innervation

In males, the sensory innervation of the genitalia is of greatest density in the glans penis, relative to the prepuce, penile shaft, and scrotal skin. Significant species differences exist in genital receptor distribution, but various encapsulated endings are typically present, including the genital corpuscles and Pacinian corpuscles, in addition to free endings. Physiologically identified receptors in the glans, prepuce, and penile shaft include slowly and rapidly adapting mechanoreceptors with small receptive fields and high rates of displacement-elicited discharge. Specific warm and cold receptors have also been found. Aspects of genital sensory function that remain to be elucidated include the effect of erection on receptor function and the significance of prominent species differences in penile morphology. Penile afferents arise in the dorsal nerves of the penis, which join with afferents from the scrotum to run in the pudendal nerves. These afferents then diverge, entering the dorsal roots at several spinal segments, primarily S2 in most species (L6 in rats). Pudendal afferents also ascend the dorsal columns to terminate in nucleus gracilis.

In females, genital sensory innervation is most dense in external structures, including the clitoris, the labia, and vaginal vestibule (corresponding to the urogenital sinus in nonprimates). The vagina is least innervated, except for a greater afferent supply to the cervix. The female external genitalia contain the same morphological receptor types as the male, with the greatest receptor density in the clitoris. The vaginal tract and cervix are innervated by free endings and in some species, especially cats, variously sized encapsulated receptors. Mechanoreceptors in the clitoris, urogenital sinus, and adjacent perivulvar tissues have very restricted receptive fields and rapidly adapting responses. Vaginal afferents, however, exhibit tonic responses to stretch of the vaginal tract or to pressure on the cervix.

As in males, afferents from the female external genitalia run through the pudendal nerves. These afferents arise from the paired dorsal nerves of the clitoris (the homologs of the dorsal nerves of the penis) and auxiliary branches from adjacent tissues. These fibers enter the upper sacral dorsal roots, primarily through S2 in most species. Most vaginal afferents, myelinated and unmyelinated, ascend to the cord through the hypogastric nerves, entering at the lowest two thoracic and highest two lumbar segments. A few vaginal afferents run in the pelvic nerves, entering through S2 and adjacent dorsal roots. Some vaginal afferents enter the spinal cord through the ventral roots.

In both sexes, the perineal musculature receives an extensive innervation from myelinated afferents, some of which arise from spindles. These muscle afferents run through the sacral plexus via several nerves, including the pudendal.

Spinal genital sensory neurons

Spinal dorsal horn neurons with genital cutaneous receptive fields lie principally in the upper sacral segments. These neurons show rapidly adapting mechanoreceptive properties and have receptive fields that are small and restricted to the ipsilateral genital structures, or extensive, including adjacent extragenital skin regions. Spinal neurons responsive to vaginal or cervical stimulation tend to lie deeper in the dorsal horn than cells with cutaneous receptive fields. These vaginally responsive neurons typically show long-duration responses to vaginocervical stimulation and exhibit convergence of low- and high-threshold mechanoreceptive afferents from cutaneous and visceral sources.

Genital sensory neurons in the brain

The male and female external genitalia are somatotopically represented in coordination with other body regions in the subcortical targets of the dorsal column-medial lemniscus and spinothalamic systems. These subcortical representations are located in nucleus gracilis of the medulla and in the most lateral region of the ventroposterolateral thalamic nuclei (VPL). Cortically, the genitalia are principally represented in the first somatosensory projection zone (on the dorsomedial postcentral gyrus of primates) and also in the second somatosensory cortical region. The cutaneous receptive fields of genital sensory neurons in the thalamic and cortical somatosensory structures often include lumbar dermatomal regions, but neurons with small receptive fields confined to the external genitals also exist. These lemniscal neurons display typically rapidly adapting tactile responses. Vaginal stimulation activates neurons within the sacral projection zone of VPL, some of which have tactile receptive fields including the external genitalia. The cervix, however, is represented in the trunk projection zone of VPL and in the first somatosensory cortical region, as well as in the orbital cortex.

Apart from the somatotopically organized lemniscal system, there is an extensive extralemniscal system of neurons that responds to genital stimuli. Males have received little investigation, so only females can be considered. Microelectrode studies in cats, rats, and monkeys have demonstrated that vaginal stimulation evokes responses in neurons that are distributed widely in the brain stem reticular formation, in some cranial nerve motor nuclei, the central gray, and the tectum. These

Figure 1. Examples of brain stem neuronal responses to vaginal probing in an anesthetized squirrel monkey. In A and B, the top traces show an integrated record of the discharges of a single neuron. The bottom traces show a pressure transducer signal which indicates the vaginal probe stimulus. A. Activity recorded from a pontine reticular formation neuron, where the first application of probing initially contacted the cervix, eliciting a pronounced acceleration of firing. The second application of probing was initially shallow and the cell failed to respond until more forceful thrusts contacted the cervix. B. An accelerative response of a lateral medullary reticular nucleus neuron to forcefully pinching the skin of a foot is suppressed by pressure on the cervix. At the left, application of the pinch (bracket) elicits an acceleration of firing that is suppressed by pressure of the probe against the cervix. The response to pinch resumed at the termination of cervix pressure. At the right, cervix pressure preceded and outlasted the pinch and limited the pinch-evoked response to a weak acceleration.

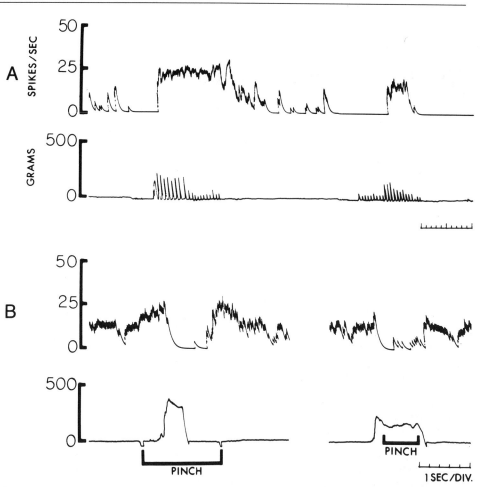

brain stem genital sensory neurons usually show several salient response properties: (1) pronounced, rapid changes in firing that often become enhanced with stimulus repetition; (2) usually stronger response to cervix than vaginal tract stimulation; (3) long poststimulus response duration; (4) diverse response patterns; and (5) convergent responsiveness to innocuous or nociceptive mechanical stimulation of visceral and extragenital cutaneous regions. Figure 1 illustrates several of these response characteristics. Vaginally responsive neurons are also found in the medial and lateral hypothalamus, the subthalamus, the medial, intralaminar and posterior thalamic nuclei, and in limbic regions (including the septal and amygdaloid nuclei and cingulate gyrus). Responses of limbic and hypothalamic neurons are generally longer latency and simpler in pattern than those seen in the brain stem or thalamus.

Vaginocervical stimulation can block or attenuate the responses to nociceptive stimuli of pontomedullary reticular and ventrobasal thalamic neurons in rats. A similar effect occurs in brain stem neurons in monkeys. This phenomenon has a behavioral parallel in rats, since vaginocervical stimulation appears to reduce the aversiveness of noxious stimuli. Recent psychophysical studies in humans have also demonstrated a pain-attenuating effect of vaginal stimulation.

Functions of genital sensory systems

The component of this system located in the somatosensory lemniscal structures, including the ventrobasal thalamus and associated somatic cortex, probably mediates the sensory-discriminative aspects of genital sensation. The extralemniscal component ascending through the reticular formation to nonspecific thalamic, hypothalamic, and limbic structures, is the probable generator of the motivational and affective properties of genital sensations. These unique attributes are likely to result from the specialized response dynamics, the cross-modal interactions between genital and extragenital somatic systems, and intrinsic properties of the anatomical substrates involved. In addition, the extralemniscal system is also important for neuroendocrine mechanisms regulating fertility (e.g., induced ovulation, prolactin release) and together with the lemniscal component, for control of species-typical patterns of copulatory behavior, both of which are vital to successful reproduction. In relation to reproductive mechanisms, it has been observed that systemic estrogens affect brain stem responsiveness to vaginal stimuli by increasing the number of responsive neurons in monkeys and by altering the patterns of neuronal response in cats. Neural responsiveness to cutaneous tactile stimuli, at levels from the pudendal nerve to the brain stem, is also facilitated by estrogen action.

Further reading

Cottrell DF, Iggo I, Kitchell RL (1978): Electrophysiology of the afferent innervation of the penis of the domestic ram. *J Physiol* 283:347–367

Price DD, Bushnell MC, Iadarola MJ (1981): Primary afferent and sacral dorsal horn neuron responses to vaginal probing in the cat. *Neurosci Lett* 26:67–72

Rose JD (1979): Anatomical distribution and sensory properties of brain stem and posterior diencephalic neurons responding to genital, somatosensory, and nociceptive stimulation in the squirrel monkey. *Exp Neurol* 66:169–185

Ueyama T, Mizuno N, Nomura S, Konishi A, Itoh K, Arakawa, H (1984): Central distribution of afferent and efferent components of the pudendal nerve in cat. *J Comp Neurol* 222:38–46

Gravitational Effects on Brain and Behavior

Laurence R. Young

On earth, the responses of many different sensory organs normally are combined to determine our sensation of which way is down. Visual, vestibular, tactile, proprioceptive, and perhaps auditory cues are combined with knowledge of commanded voluntary movement to produce a single, usually consistent, perception of spatial orientation. Angular stabilization of the eye to reduce retinal image slip and stabilization of head and body position with respect to the vertical to avoid falling are also based upon this multisensory integration process. When tilting one's head to the shoulder, for example, this voluntary movement is confirmed to the brain by signals from the muscle and joint receptors, from the tilted visual field, from the increased pressure on one foot, and from the two portions of the balance mechanism of the inner ear: the semicircular canals, which sense angular motion, and the otolith organs, which sense linear acceleration and gravity. In the weightless free-fall condition of orbital space flight the correspondence among the signals is drastically altered. The otolith organs no longer indicate anything meaningful concerning the static orientation of the head. The dense mass of the otoconia no longer pulls the otolithic membrane downhill, bending the hair cell cilia of the maculae when the head is tilted. Rather, like any other linear accelerometer, the output of the otolith organs in weightlessness is limited to indications of the short-duration transient linear accelerations during head movements. Once deprived of the normal static orientation information from the otolith organs, the brain must rely upon other senses to set up a reference frame with respect to which the astronaut can judge his orientation. Visual signals play an increasing role in spatial orientation in weightlessness.

The recent Spacelab flights have provided especially valuable observations on the effects of weightlessness and space flight. During the initial stages of adaptation to weightlessness, a conflict exists between the outputs of the otolith organs and the remaining senses, especially associated with voluntary head movements. This conflict is presumed to be the basis of space motion sickness, a malady which affects roughly half of all space travelers and which typically lasts two or three days. As visual cues become dominant, the astronauts begin to orient such that the surface upon which they are working becomes a vertical wall and the place where their feet touch, if they are indeed touching, becomes the floor, or the down reference. Unusual visual orientations, like seeing a fellow crew member upside down, entering a new part of the spacecraft in an unusual orientation, or looking out the window and seeing the earth at the top of the window and the sky at the bottom may prove disturbing and even bring on motion sickness symptoms. Moving visual fields create a greater sense of self motion, and otolith cues begin to be ignored as the astronaut's brain undergoes the reinterpretation of his sensory signals. The limbs no longer have any weight or require any muscle tension when static, other than what is required to overcome internal elastic-

ity. Knowledge of limb position when the muscles are relaxed may be degraded, and the astronaut is occasionally unaware of limb position, which tends to be more flexed than in 1 g. The ability to estimate the mass of objects, in the absence of their weight, is reduced. As measured by the Hoffman reflex during transient accelerations from weightlessness, spinal cord excitability may be greatly reduced. Ocular stabilization during head movements may be impaired, especially for the nodding and tilting motions that normally involve otolith system contributions.

In space, the otolith organs respond only to linear acceleration, as the brain may reinterpret their signals to represent only translation rather than tilt. On earth, such an interpretation would result in the wrong signals to the postural control system to prevent falling, or to reduction in the small compensatory torsional motions of the eye when tilted. Following return to earth astronauts may exhibit wide stance and unsteady gait, and difficulty balancing on a narrow rail with eyes closed. (Reduced blood supply to the brain, associated with cardiovascular deconditioning and blood pooling in the legs when standing just after reentry, may contribute to the unsteadiness.) Head movements may result in unusual illusions of self motion or of ground movement, as the otolith organ signals and perhaps the joint receptors become recalibrated to the terrestrial environment. Return to normal earth functioning may take from one or two days following short flights to several days or weeks for long flights. Part of the recovery is associated with rebuilding the leg muscles, which tend to atrophy from disuse in space. Part of the readaptation is also in the brain, which must reinterpret the sensory cues appropriate for earth.

Animal experiments concerned with brain function in weightlessness have been limited. Monkeys can become motion sick, and they may show altered eye movement patterns. Goldfish, when deprived of gravito-inertial orientation forces on their graviceptors, will begin a series of looping swimming motions that may serve to satisfy a drive to remain upright. Spider webs lose their normal regularity.

The sites of adaptation to weightlessness have not yet been determined. Like many other examples of plasticity to sensorimotor rearrangement such as the wearing of reversing prisms, the adaptation is probably central. Other theories involve the end organs themselves. Space experiments can investigate changes in their morphology, such as the number and size of otoconia, or changes in the mechanics of the transducers when the steady load of 1 g is removed. Preliminary results concerning otolith morphology in the rat and otolith organ afferent responses in the frog are inconclusive.

Further reading

Graybiel A (1973): The vestibular system. In: *Bioastronautics Data Book*, Parker J, West V, eds. Washington DC: National Aeronautics and Space Administration, SP-3006

Nicogossian AE, Parker JE (1982): *Space Physiology and Medicine*. Washington DC: National Aeronautics and Space Administration, SP-447

Special Issue on Spacelab 1 (1984): *Science* 225 (4658)

Young LR (1984); Perception of the body in space: Mechanisms. In: *Handbook of Physiology—The Nervous System Vol 3, Sensory Processes, Part 1*, Bethesda: American Physiological Society

Special Issue on Spacelab D-1 *Experimental Brain Research* Oct. 1986

Hair Cells, Sensory Transduction

A.J. Hudspeth

Hair cells are the sensory receptors in the organs of the vertebrate internal ear and in the lateral-line organ. They provide sensitivity to the broad range of stimuli to which these acousticolateralis organs are responsive: air- and water-borne sound, substrate vibration, water motion, and angular and linear acceleration, especially that due to gravity. Hair cells are of particular importance to humans because a sensitive vestibular apparatus is required for our upright, bipedal mode of locomotion and because a sense of hearing is of paramount importance in our verbal communication.

Hair cells arise from the bilateral otic placodes, disks of superficial ectoderm that subsequently invaginate to create the otic cysts. The organs of the internal ear, or labyrinth, develop by growth and complex remodeling of the cysts; the lateral-line organs of fishes and aquatic amphibians originate from cells deposited along the body surface by blast cells migrating from the placodes. The ganglion cells that innervate hair cells in the internal ear, whose fibers largely constitute the eighth cranial nerve, are also at least partially of placodal origin.

As a consequence of their ectodermal origin, hair cells are epithelial. Most are cylindrical or flask-shaped cells ligated to the contiguous supporting cells by a junctional complex; tight junctions in this complex separate the dissimilar fluids above and below the hair cells. Although they lack axons and dendrites, hair cells nonetheless make afferent synapses and receive efferent ones upon their basolateral surfaces.

The organelle that gives the hair cell its name, the hair bundle, is a mechanosensitive structure at the cellular apex (Fig. 1). The hair bundle comprises from 30 to 300 processes, all but one of which are of similar structure. These stereocilia are hypertrophied microvilli, each consisting of a sheath of plasmalemma enveloping a fascicle of cross-linked actin microfilaments. Stereocilia range in length from under 1 μm to over 50 μm; in general, shorter stereocilia occur in auditory organs with high-frequency sensitivity, while longer ones are typical of vestibular organs that respond to static or low-frequency stimuli. Each hair bundle also possesses a single, eccentrically located kinocilium, a true cilium with an axonemal (9 + 2) array of microtubules.

The arrangement of the processes within each hair bundle is strikingly regular. The stereocilia in every instance reside in a hexagonal array. Moreover, they always vary in length monotonically across the hair bundle. A hair bundle is accordingly not flat across its top, but is instead beveled like a hypodermic needle. Because the direction of its taper corresponds with one of the hexagonal axes along which the stereocilia are disposed, the hair bundle possesses a plane of bilateral symmetry. The kinocilium lies at the tall edge of the bundle, adjacent to the longest stereocilia; in certain organs such as the mammalian cochlea, however, the kinocilium degenerates at or around the time of birth.

Despite the range of stimuli to which acousticolateralis organs are responsive, every vertebrate hair cell appears to operate in the same way. Sounds, vibrations, water motions, or accelerations act through various mechanical linkages to exert a force upon the distal end of the hair bundle. This mechanical input causes each stereocilium to pivot near its base, producing a deflection of the bundle's distal tip that ranges in amplitude from about 100 pm at threshold to over 1 μm at the point of trauma. The mechanical stimulus modulates the opening of transduction channels that occur in the hair bundle, probably near the distal tips of the stereocilia. A hair cell's mechanosensitivity is strikingly vectorial: moving the bundle's tip toward the tall edge of the hair bundle (and the kinocilium) opens transduction channels and depolarizes the cell, moving the bundle in the opposite direction closes channels and hyperpolarizes the cell, while motions at a right angle to the cell's plane of morphological symmetry are without effect. While the basis of mechanoelectrical transduction has not been established, it appears possible that channel molecules are gated by shearing motions between the distal portions of contiguous stereocilia.

Two features of the transduction channel of hair cells are noteworthy. First, gating of the channel is remarkably rapid, so that the latency of responses measured in vitro is only a few microseconds. This finding suggests that mechanical stimuli act directly upon transduction channels, without the intervention of second messengers. The great rapidity with which transduction occurs in hair cells accounts for our capacity to hear sounds at frequencies as high as 20 kHz, and for bats and whales to discern stimuli at 100–200 kHz. The second interesting property of the transduction channel is that it is relatively nonselective in its ionic permeability. At least in the organs of the internal ear, transduction current normally is carried primarily by K^+, which occurs at a high concentration in the fluid, endolymph, bathing the apical surfaces of hair cells. In vitro experiments indicate, however, that other alkali cations carry transduction current equally well; divalent ions such as Ca^{++} and small organic cations such as tetramethylammonium also permeate the transduction channel. On the basis of its ionic permeability, the transduction channel probably possesses a hydrated pore at least 700 pm in diameter. The poor selectivity of the channel may account for the susceptibility of transduction to numerous ototoxins, particularly antibiotics of the aminoglycoside family (streptomycin, gentamicin, etc.). These substances appear to exert a reversible, voltage-dependent block on transduction by occluding the pore.

In addition to transducing mechanical stimuli into electrical signals and synaptically transmitting information to eighth-nerve fibers, hair cells contribute to the frequency selectivity of many organs in the internal ear. Two modes of tuning have been demonstrated. In some organs, the physical proper-

Figure 1. The hair bundle and apical cellular surface of a vertebrate hair cell. This bundle includes about 50 stereocilia and a single kinocilium, the process culminating in a bulb. The stereocilia occur in a hexagonal array and systematically increase in length from the left to the right edge of the bundle; stimulation in the same direction evokes a depolarizing receptor potential. In this scanning electron micrograph of a cell from the sacculus of the bullfrog's internal ear, the hair bundle is about 8 μm tall and the constituent stereocilia are approximately 0.4 μm in diameter. The hair cell is surrounded by supporting cells endowed with ordinary, short microvilli.

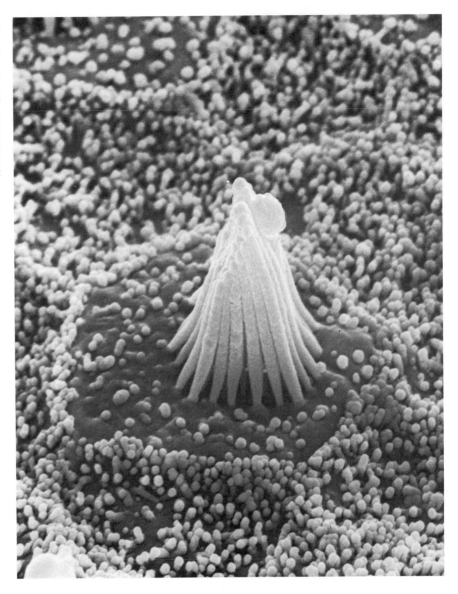

ties (mass, elasticity, and hydrodynamic drag) of a hair bundle interact to make it mechanically resonant at a certain frequency. In other instances, the interplay of ionic conductances in the hair cell's membrane produces an electrical resonance that tunes the cell to a particular frequency. It is probable that mechanical and electrical tuning processes coexist in some cells; still more stimulating is the possibility that some hair cells, such as the outer hair cells of the mammalian cochlea, are mechanically active and contribute to tuning through electromechanical feedback.

Further reading

Corey DP, Hudspeth AJ (1983): Kinetics of the receptor current in bullfrog saccular hair cells. *J Neurosci* 3:962–976

Crawford AC, Fettiplace R (1981): An electrical tuning mechanism in turtle cochlear hair cells. *J Physiol* (*Lond*) 312:377–412

Hudspeth AJ (1983): Mechano-electrical transduction by hair cells in the acousticolateralis sensory system. *Annu Rev Neurosci* 6:187–215

Hudspeth AJ (1983): Transduction and tuning by vertebrate hair cells. *Trends Neurosci* 6:366–369

Headache

F. Clifford Rose

Headache should mean an ache in the head, but surprisingly there is disagreement about the precise meaning of both these terms. The Oxford dictionary defines headache as "a continuous pain in the cranial region of the head," which is not clinically acceptable since headache need not be continuous, nor indeed be a pain. The term cranial, strictly anatomically, is applied to the bones covering the brain and is presumably used in this sense in the Oxford dictionary, but, to the clinician, cranial applies to the bones of the whole head, i.e., the skull. This is not semantic sophistry, since it is relevant to the question of whether pains in the face should be included under the title of headache. The definition of head in the same dictionary is the "anterior part of the body of an animal . . . , it contains the brain and the special sense organs and the mouth," indicating that the term should include that region above the neck, a meaning widely accepted.

The brain itself does not contain sensory nerve endings that subserve pain, so that headache originates either from the blood vessels or the coverings of the brain (meninges). Of the three fossae of the cranial base, lesions in the anterior and middle fossae give frontal headache, while those in the posterior fossa may be felt in the back of the head or neck. Pain in the head can be referred from disorders of the eyes, sinuses, teeth, and neck, but these are not common. There are many types of headache, but the two most common are migraine and muscle contraction headache.

Muscle-contraction headache

Although the terms muscle-contraction headache, tension headache, and psychogenic headache have been used interchangeably, attempts have been made to distinguish them. This term is inaccurate in describing pathogenesis, since it is unlikely that the pain is caused by contraction of the scalp muscles, not least because this is no greater than in normal controls. Nor is there any evidence to confirm the hypothesis of ischemia caused by constriction of the vessels by the muscle contraction. The term tension headache is equally confusing because it has been variously interpreted as meaning tension of the muscles rather than psychological tension, and the pathogenesis of the pain is still unexplained.

When the symptoms are acute, stress factors are usually elicited, but in the chronic form, they are less easily determined. The headache is usually bilateral and described as dull, tight, pressing, and constant. More commonly than not, there is an associated depression, and in these cases relief is often obtained by antidepressant drugs, a therapeutic strategy that may also work when depression is not a feature, supporting the hypothesis that the pain may be central in origin and perhaps due to depletion of amines in the endogenous pain control system. That headache can be associated with florid psycholog-ical disturbances is indisputable, not least because it disappears when mental health is restored.

Combined, mixed or tension-vascular headache

In approximately one-third of patients, clinicians have difficulty in assigning the appropriate diagnosis, often because symptoms are present that are found in more than one diagnosis. The chief diagnostic difficulty has been between common migraine and muscle contraction headache. When characteristics of both are present, and this is not uncommon, the terms combined, mixed, or tension-vascular headaches have been used.

Although this type of headache is well recognized clinically, it is probably neither vascular nor due to muscle contraction. It is frequently argued that common migraine, tension-vascular headaches, and muscle contraction headaches are all part of the same syndrome, a continuous clinical spectrum at one end of which is muscle contraction headache and at the other end is migraine.

Epidemiological and statistical analyses suggest that which of these diagnoses is made is related to the severity and frequency of the headache. Increasing severity of headache was associated with an increasing number of symptoms that characterized migraine, namely, unilaterality, gastroenterological features, and "warning." As headaches become more frequent, symptoms that characterize migraine decrease in frequency.

Cluster headache

Cluster headache is a distinct clinical entity separate from migraine. A wide variety of terms have been used (see Table 1), but the two most common are cluster headache and migrainous neuralgia. Since the condition is neither migrainous nor neuralgic, most authorities use the term cluster headache since it describes the essential clinical characteristic of a period during which the short, sharp attacks occur.

Table 1. Synonyms for Cluster Headache

1. Hemicrania angioparalytica (Eulenberg 1878)
2. Ciliary neuralgia (Harris 1926)
3. Periodic migrainous neuralgia (Harris 1936)
4. Erythromelalgia of the head (Horton et al. 1939)
5. Histamine cephalagia (Horton 1941)
6. Horton's headache
7. Greater superficial petrosal neuralgia (Gardner et al. 1947)
8. A particular variety of headache (Symonds 1956)
9. Sphenopalatine neuralgia (Sluder 1908)
10. Erythrosopalgia (Bing 1913)
11. Vidian neuralgia (Vail 1932)
12. Autonomic cephalalgia (Byrickner and Riley 1935)

One well known definition of cluster headache is "It consists of recurrent attacks of unilateral intense pain involving the eye and head on one side, usually associated with unilateral nasal congestion and rhinorrhoea and unilateral lacrimation and redness of the eye. They occur one or more times daily and last for 10–240 minutes. The bouts continue for weeks or months but may be chronic." In each cluster, the individual attack nearly always keeps to the same side, but it is not uncommon for the sides to change in different cluster periods. The severe intensity of the pain is characteristic but not invariable. Horner's syndrome occurs in about 20% of cases during the attack and occasionally persists.

In the chronic form, the patient has at least two attacks weekly for at least one year. In the secondary chronic form, the patient begins with episodic cluster periods that then become continuous. In the primary chronic form, from the onset there are no episodes and the disorder persists.

Food

This has often been invoked as a provocative factor, e.g., "ice cream headache," where frontal headache is caused by the excessive stimulation of swallowing ice-cold substances. The "Chinese restaurant syndrome" is due to monosodium glutamate, a vasodilator, and "hot dog headache" is caused by other vasodilators, such as nitrites, used in curing meat.

Further reading

Rose FC, ed (1984): *Progress in Migraine Research 2 (Progress in Neurology Series)*. London: Pitman Books

Rose FC, ed (1985): *Handbook of Clinical Neurology Vol 4*. Vinken P, Bruyn G, eds. Amsterdam: North Holland Elsevier

Rose FC, ed (1985): *Migraine: Clinical and Research Advances*. Basel: Karger

Infrared Sense

Peter H. Hartline

Infrared (IR) radiation is electromagnetic energy that has the same form as visible light but has wavelength greater than about 700 nm and is not detected to any appreciable degree by photoreceptors. Natural objects radiate such energy approximately in proportion to the fourth power of their absolute temperature in accordance with the Stephan Boltzman law of black body radiation. Animals that are warmer or colder than their surroundings can be thought of as infrared (3–20 μm wavelength) luminous sources that are respectively brighter or darker than the background against which they are observed. Reptiles (and perhaps bats and insects) have evolved sense organs that are responsive to infrared radiation of the intensities radiated by objects and animals in the natural world, and thus may mediate a specialized infrared sense. Mammals, including primates, have skin sensory receptors that respond to warmth and thus to infrared radiation; however, their sensitivity is such that only objects that are very warm (such as fire or the sun) or nearby are detected via their IR radiation. Such receptors are discussed under *Thermoreceptors*.

Noble and Schmidt obtained the first evidence of infrared sensitivity in a vertebrate by their behavioral studies of rattlesnakes. Blindfolded snakes struck at warm lamp bulbs that were covered with black hoods, but lost this ability if their pit organs were occluded. Subsequent experiments have demonstrated that rattlesnakes can strike toward warm objects with 2–3 degrees resolution on the basis of their infrared sense. Vampire bats can be trained to discriminate a warm target from a cold one, but proof that they use an infrared sense in their normal behavior is lacking. Insects, notably mosquitoes, probably localize warm-blooded hosts by radiant or convective warmth, possibly via an infrared sense.

Infrared sense organ

Two taxa of snakes (subfamily Crotalinae and family Boidae) have evolved specialized infrared senses. Their sense organs consist of depressions (or pits) in the facial or labial scales, at the bottom of which are trigeminal nerve fiber terminals. Crotaline snakes (so-called "pit vipers") have a single pit on each side of the face. A thin sheet of epithelium ("pit membrane") is suspended in air, near the base of the pit; it contains about 5000 heat-sensing terminals. Each terminal contains extensive ramifications of a trigeminal sensory axon embedded in Schwann cell elements and covering an area 40–60 μm in diameter, and overlapping neighboring terminals little if at all. The pit's optics resemble that of a pinhole camera whose (lens-less) opening and depth measure a few millimeters. Boid snakes (pythons, boas) have several pits along the upper and lower lips on each side of the head. Nerve endings occur at depths that are unusually shallow, considering cutaneous receptors. The designs of both types of organs facilitate rapid change of temperature of the nerve endings in response to changes in thermal flux; furthermore, a different restricted patch of receptors is illuminated depending on the angular location of an infrared source, which forms the basis for spatial analysis of infrared radiation emanating from objects or animals in the snake's surroundings.

Vampire bats (*Desmodus rotundus*) have a series of perinasal depressions that is strongly reminiscent of the labial pits of boid snakes and that may be infrared sense organs. The trigeminal nerve that innervates the face contains some warm fibers, but thus far these have been found to innervate primarily the nose-leaf, not the perinasal depressions.

In mosquitoes (diptera), bees (hymenoptera), and beetles (coleoptera), sensillae of defined morphology are associated with afferent fibers that are sensitive to warming. These sensory structures are often characterized by lamellar in addition to conventional dendritic endings in sensillae without pores. The superficial innervation, large surface area, and low total thermal capacity of the sense organ is conducive to sensitive response to infrared radiation, though this also favors sensitivity to ambient air temperature. Other arthropods have sensillae morphologically similar to those with demonstrated warm-receptive function. Examples are found in moths (lepidoptera), cockroaches (orthoptera), aphids (homoptera), and spring-tails (collembola).

Transduction and receptor sensitivity

T.H. Bullock and collaborators first proved that snake pit organs are sensitive to infrared radiation. Their work demonstrated that: (1) the afferents exhibit "background" nerve impulse activity in the absence of spatially or temporally patterned infrared stimulation; (2) increase of infrared radiant flux incident on the receptors causes a transient (1–2 sec) increase of the frequency of nerve impulses in trigeminal pit afferents; (3) following the transient, impulse activity has a higher mean frequency than the initial average background firing rate, but this decays slowly toward background over a time-course of many minutes; (4) a similarly transient depression of activity is caused by decrease of incident infrared radiation. Thus, change in infrared flux appears to be signaled by a frequency code in primary afferents. Experiments in which the time course of temperature rise in the pits was controlled or estimated indicate that the time derivative of temperature is the parameter most predictive of response magnitude. A temperature rise less than 0.003°C in 0.06 sec is suprathreshold.

The biophysical mechanisms of transduction are unknown. Quasi-intracellular records from microelectrodes inserted into the terminal region suggest that a tonic depolarization is caused by warming; the ionic mechanism is unknown. The action spectrum for primary afferents does not indicate the presence of spectral specialization. Electron microscopic evidence indicates that, in warmed but not cooled terminals, mitochondria have morphology characteristic of those in cells with active sodium pumps, but it is not known whether this is the cause or effect of the generator current. The involvement of a specific photochemical transductive mechanism operating in the far infrared is regarded as unlikely. There is general acceptance that infrared radiation causes a temperature rise that is transduced to neural signals by thermoreceptive warm receptors.

The sensitivity of the pit's warm fibers to temperature increase (a rate of change of 0.03–0.06°C/sec is above threshold) is not extraordinary among vertebrate thermal senses. But the low heat capacity and superficial location of nerve endings confers much greater sensitivity to infrared radiation and much more rapid responses in boid and crotaline snakes than is reported for mammals. Thus the approximate threshold for response to trigeminal fibers is an increase of infrared flux of $5–8 \times 10^{-5}$ W/cm^2 incident on the rattlesnake pit membrane, about what is radiated to the pit by a human hand or a rat at about 50 cm distance. Most primary afferents respond as if they receive excitation from one punctuate afferent terminal; there has been no report of inhibitory interaction at the level of the trigeminal nerve. Excitatory receptive fields are usually 50–70 degrees across. Boid afferents are about 25% as sensitive as those of Crotalinae. In neurons of the rattlesnake central nervous system, sensitivity to change of flux is about a factor of 10 greater than that of primary afferents.

Little physiological work has been done on neural responses of the candidate infrared organs of bats. Vampire bats have been reported to learn to discriminate a difference of infrared flux of 5×10^{-5} W/cm^2. Warm fibers innervating the nose-leaf and pads of these bats are excited by large temperature steps, but their sensitivity appears insufficient to explain the behavioral responses.

Insect warm fibers are usually found in association with cold fibers and hygroreceptive fibers (innervating the same sensillum). They exhibit transient and tonic excitation to onset of infrared radiation or to flow of warmed air. The transient response probably reflects sensitivity to the rate of change of temperature. Sudden air temperature change as small as 0.05°C can yield a detectable increase in frequency of discharge, but the induced change of temperature of the sensillum has not been measured or estimated, and sensitivity to infrared radiation has not been measured.

Central sensory pathway and neural integration

Nothing has yet been reported about central integration of infrared sensory information in vampire bats or insects. In crotaline and boid snakes, pit afferents project to the nucleus of the lateral descending trigeminal tract (LTTD). Postsynaptic neurons, in rattlesnakes, may have excitatory receptive fields about half as broad as those of primary afferents; many have inhibitory regions flanking their excitatory receptive fields. In rattlesnakes, horseradish peroxidase (HRP) and degeneration studies have shown that the LTTD sends axons to a homo-

lateral nucleus in the medullar reticular formation, nucleus reticularis caloris (RC); this nucleus appears to be a subdivision of LTTD in boidae. Anterograde and retrograde HRP studies have shown that the RC in turn projects to the contralateral optic tectum. Tectal neurons are particularly responsive to moving, warm stimuli. Their receptive fields span the size range of primary and LTTD units and frequently have inhibitory regions. Thus a specialized and segregated subdivision of the trigeminal sensory system has evolved, which brings information about infrared stimuli to the tectum.

The vertebrate optic tectum (termed superior colliculus in mammals) receives highly ordered visual input in which adjacent spatial locations are represented by responsive neurons in adjacent tectal locations. The result is a spatiotopically organized map of the visual field. The infrared systems of boidae and crotalinae similarly form spatiotopically organized maps of the field of view of the pit organ. Infrared-responsive neurons lie in the tectal layers beneath the superficial ones that receive visual input. However, the visual and infrared maps correspond approximately to each other in such a way that visual and infrared neurons at the same tectal location (in their corresponding layers) respond to visual or infrared stimuli (respectively) from approximately the same spatial location, though some systematic discrepancies have been noted.

The optic tectum is a brain center that receives information about spatial location of objects, and is involved with initiating behaviors that are oriented toward appropriate spatial locations. Thus the infrared system is favorably connected to permit orientation toward appropriate patterns of infrared radiation. It is reasonable to suppose that the behavioral responses that these snakes exhibit toward warm stimuli can be at least partly accounted for by the known trigeminotectal pathway.

Whether more complex stimulus recognition tasks such as shape discrimination can be mediated by the infrared system is not known. Such sensory capabilities are thought to be mediated by forebrain structures (sensory cortex) in mammals. There is anatomical and electrophysiological evidence that infrared information ascends through the thalamus (n. rotundus) to the forebrain (dorsal ventricular ridge) in crotaline snakes. However, little is known of the response properties of neurons in these regions, and their behavioral significance is not known at all.

Further reading

Hartline PH (1974): Thermoreception in snakes. In: *Handbook of Sensory Physiology* III/3: Electroreceptors and other specialized receptors in lower vertebrates. Fessard A, ed. New York: Springer-Verlag

Molenaar GG (1987): Anatomy and physiology of infrared sensitivity of snakes. In: *Biology of the Reptilia* Vol 17, Neurology C Gans C and Northcutt RG, eds. New York, London: Academic Press (in press)

Newman EA, Hartline PH (1982): The infrared "vision" of snakes. *Sci Am* 246(3):116–127

Insect Communication, Intraspecific

Franz Huber

Intraspecific communication involves the emission of species-specific signals by one member (the sender), their reception and processing by a second member (the receiver), and an adaptive effect on the subsequent behavior of both organisms. Insects exhibit intraspecific communication strategies during reproductive as well as social behaviors, such as parental, agonistic, and group spacing tactics. Their communication deals with signal emission and reception in the chemical, visual, auditory, and vibratory modalities, and quite often several modalities are involved. Signals in some of these modalities deteriorate somewhat in traveling through the environment but keep their species-specific context.

The sender-receiver relationship can be studied in two ways: (1) by a quantitative measure of the input-output function in the behaving animal under controlled conditions (behavioral approach), and (2) by analyzing sensory, neuronal, and effector mechanisms at the single-cell or multicellular level (neurobiological approach). The interdisciplinary study of intraspecific communication must incorporate concepts, methods, and results within the field of ethology (animal behavior), and it must be based on concepts, methods, and results obtained in the whole field of neuroscience, including molecular, cellular, and multicellular levels. Several disciplines contribute to communication studies, for instance, behavioral ecology, sociobiology, sensory physiology, neurophysiology, information theory, and control systems analysis.

By comparing the results obtained in the two approaches a correlation may be found between behavior and cellular events, for instance, in the strength of responses, their time course, or in other parameters. However, such a correlation does not necessarily indicate a causal relationship.

It is not my intention to cover the manifold strategies of intraspecific communication in the insect world; instead, one example is outlined to guide the reader to present-day knowledge in this field and to elucidate further research work.

The communication strategy most familiar to me and perhaps the one best studied in the past in both the behavioral and neurobiological fields involves emission, reception, and processing of airborne sound signals in crickets. These insects, represented by about 2000 different species spaced all over the world, start their reproductive behavior with the calling song of the adult male. This signal, characteristic for each species, attracts those adult females in the state of sexual responsiveness. The female approaches the male by walking or flight, and this strategy is called phonotaxis.

To understand signal production by the male as the sender, knowledge is required about the morphological structures (frontwings), their mode of operation (biomechanics), the muscles driving the wings (electromyogram recordings of muscles in unrestrained singing males), the motoneurons driving the muscles, and interneurons that are connected with motoneurons and higher centers, such as the brain, that steer and control this behavioral display called stridulation. In addition, one needs to know how exteroceptive and proprioceptive sensory feedback comes into play to stabilize the signal pattern in the frequency, intensity, and time domain.

The receiver, the female, has to solve two equally important tasks. She must recognize the species-specific message and discriminate it from other messages in the same modality range (recognition function), and furthermore, she must localize the position of the sender in space (orientation function).

A behavioral paradigm was developed in recent years, the walking compensator (a spherical treadmill), where the female can exhibit phonotaxis in an open loop system. It was found that the female requires at least two parameters of the species-specific calling song: the carrier frequency band (4–5 kHz) and a time window of syllable intervals ranging from about 25 to 55 msec (Fig. 1A).

This range covers the natural range of frequencies and syllable intervals of the calling male.

In the search for sensory and neuronal correlates it was first found that the receptors in the cricket ear, which is placed in the frontleg, are tuned (most sensitive) to a variety of frequencies, inside and outside the intraspecific communication range. Those receptor cells, tuned to the calling song carrier band, fulfill the requirement in the frequency domain (parameter 1). They encode intensity of the sound signal by spike frequency (log relationship), but they are not tuned to the species-specific temporal pattern of chirps and syllables in the song. Thus, they cannot be considered as an intimate part of the recognizer device.

With intracellular recording and staining techniques (using fluorescent dye application and consecutive histology), a family of auditory interneurons was encountered within the prothoracic ganglion. This ganglion represents the first central station where acoustic information provided from auditory sense cells is transmitted to higher order neurons. It is this ganglion where binaural information processing starts, because it contains neurons which have access to both left and right-ear auditory input.

A neural correlate was found for binaural information processing in the prothoracic ganglion as a first step toward an understanding of phonotactic tracking. This correlate is based on a mirror image pair of intraganglionic (local) interneurons with reciprocal inhibition. With the newly developed technique, the legphones, it was possible to stimulate each of the two frontleg ears separately and in combination. Intracellular recordings from the two-cell network clearly showed that each member of the pair receives excitatory input only from one ear, and it is inhibited by input affecting the other ear through its partner cell. This reciprocal inhibition sharpens binaural contrast. The information from these cells is transmitted to other auditory neurons. Among them are those which conduct information to the cricket brain. There, we assume a final

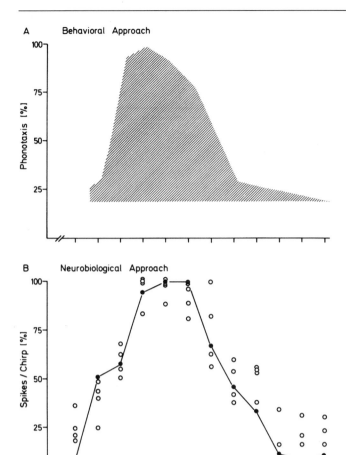

Figure 1. Temporal pattern (syllable interval) recognition of female crickets as studied with the behavioral approach (A) and neuronal correlates to temporal recognition as studied with the single-cell approach (B). A. Range (hatched area) of phonotactic tracking on the walking compensator (spherical treadmill). Ordinate, % of tracking evaluated as proportion of accurate tracking per stimulus time; abscissa, range of syllable intervals in msec. Tracking above 60% is seen in the range between 26 and 58 msec syllable intervals. This range covers the natural range of intervals in the conspecific male's calling song. B. Class 2 local brain neuron responses (spikes per chirp) with band-pass filter properties in the temporal domain. Ordinate, % of response; abscissa, as in A. The response curve, drawn for a single neuron of this class (black circles), mimics rather accurately the response curve of phonotactic tracking. Open circles are response measures from other neurons within this class, each from a different animal. Modified from Schildberger 1984.

comparator cell or network for binaural messages, the output of which may control the female's phonotactic course. Such a comparator could in principle work by calculating intensity as well as latency differences of the arriving binaural information.

Since it was known that auditory receptor cells are not tuned to the species-specific pattern of the calling song, an attempt was made to study temporal tuning and recognition within neurons of the prothoracic ganglion and the brain. Neurons within the prothoracic ganglion, some identified as intraganglionic, some as ascending, and some as descending, cannot be considered as an intimate part of the recognizer network because they copy a variety of calling songs with temporal structures inside and outside the effective phonotactic range. Within the group of plurisegmental ascending neurons, there

exists again a mirror image pair, bilaterally present, with the axon terminating in the cricket brain. The members of this pair carry information in the correct carrier frequency range (4–5 kHz), and they copy the chirp and syllable pattern with high fidelity, but they are not specifically tuned to the conspecific pattern.

Within the brain the axon terminals of the ascending neuron overlap with one class of local auditory brain neurons. This class is again tuned to the calling song carrier frequency range, but less sensitive to sound intensity when compared with ascending neurons. There is anatomical and physiological evidence that this class receives excitatory input from the ascending fibers. A second class of local brain neurons was identified with a dendritic field of arborizations overlapping those of the first class of brain neurons but without overlapping the axon terminals of the ascending neurons. This second class has specific properties. Some members of the anatomically similar group of cells act as high-pass filters. They are strongly activated by artificial songs with short syllable intervals (high syllable repetition rates) which partly lie outside the effective range. Other members of the same group act as low-pass filters. They are strongly activated with songs of long syllable intervals (low syllable repetition rates) which also partly lie outside the effective range. A third group within this class of cells has band-pass characteristics. The response of the cell is independent of sound intensity (at least in the moderate and upper range), and activation occurs only within a time window covering those syllable intervals that were shown to be effective in phonotaxis (Fig. 1B). The present hypothesis is that high- and low-pass cells converge on band-pass cells and act as an AND gate. How these cells process the input and how they convey their output to neural systems responsible for phonotactic walking or flight remains to be shown. It should be added that such band-pass neurons were recently found in the frog brain where they also could be correlated to intraspecific sound communication.

Thus it seems that insects are suitable to investigate intraspecific communication strategies at both the behavioral and cellular level. From here future steps can be undertaken, down to the synaptic properties of the cells involved and the transmitters used and to the development of the auditory pathway in both morphological and physiological terms. Steps can also be taken to touch evolutionary aspects, biotop-dependent tactics, and to work out the adaptive significance. In addition, the approaches outlined here can also be applied to communication systems using other sensory modalities and modality combinations.

Further reading

Reviews and symposia

Blum MS, Blum NA (1979): *Sexual Selection and Reproductive Competition in Insects.* New York: Academic Press

Kalmring K, Elsner N (1985): *Acoustic and Vibrational Communication in Insects.* Berlin and Hamburg: Verlag Paul Parey

Lewis B (1983): *Bioacoustics: A Comparative Approach.* New York: Academic Press

Thornhill B, Alcock J (1983): *The Evolution of Insect Mating Systems.* Cambridge: Harvard University Press

Original papers

Huber F (1983): Neural correlates of Orthopteran and Cicada phonotaxis. In: *Neuroethology and Behavioral Physiology,* Huber F, Markl H, eds., pp 108–135. New York: Springer-Verlag

Schildberger K (1984): Temporal selectivity of identified auditory neurons in the cricket brain. *J Comp Physiol* 155:171–185

Thorson J, Weber T, Huber F (1982): Auditory behavior of the cricket. II. Simplicity of calling-song recognition in Gryllus, and anomalous phonotaxis at abnormal carrier frequencies. *J Comp Physiol* 146:361–378

Kinesthesia, Kinesthetic Perception

D. Ian McCloskey

The "sixth sense" was what Sir Charles Bell, late last century, named the sense of the positions and actions of the limbs. In its entirety, Bell's sixth sense concerns perceived sensations about the static position or velocity of movement (whether imposed or voluntarily generated) of those parts of the body moved by skeletal muscles, together with perceived sensations about the forces generated during muscular contractions even when such contractions are isometric. The general descriptive terms used today for such sensations are kinesthetic (which despite its literal translation was coined by Bastian to describe this complex of sensations, including those in which movement is not a feature), and proprioceptive (which was used by Sherrington in a rather wider vein than here to include also vestibular sensations and inputs from muscles and joints that are not necessarily perceived, that is, sensations about which one might not be able to give a subjective report).

The classes of afferent fiber that are candidates for subserving kinesthetic sensibility are those from the skin, from the muscles and tendons, and from the joint capsules and ligaments. In the 19th century, however, various authors questioned the necessity for any afferent information at all, suggesting instead that normal kinesthetic sensations arise as a consequence of the effort to move and are derived in some way within the central nervous system (CNS) from centrifugal or motor signals—sensations of innervation as they were called by Helmholtz. It now seems that for some classes of kinesthetic sensation, centrally generated signals of this kind can be important, while others are based entirely upon afferent signals. The special case of perception of the direction in which the eyes are pointing is not considered here.

Position and movement

No sensation of movement or altered position accompanies a voluntary attempt to move part of the body which, because of obstruction or paralysis, is prevented from moving. This indicates that centrally generated signals cannot give rise to sensations of movement or of altered position.

Until the early 1970s it was widely believed by physiologists that the senses of position and movement could be entirely attributed to the discharges of sensory nerve endings in and around the capsules and ligaments of joints. This belief was based first on the apparent suitability of joint receptors for this role as revealed by electrophysiological recordings of their behavior during movements imposed on joints. There are several morphological types of receptor found in the joints, including paciniform endings, and bare nerve endings likely to be involved in nociception. Of the more specialized types of joint receptors, some were shown to change their rates of firing in proportion to the angular velocity of joint rotation and to adopt different, steady firing rates at different, fixed locations throughout the whole range of normal joint excursion.

Added to this was the apparent unsuitability of intramuscular receptors for kinesthesia. While stretch-sensitive receptors, such as muscle spindles, in muscles operating on a joint might have seemed appropriate to provide a signal from which the CNS could compute joint position or velocity of movement, these receptors could also be made to discharge by fusimotor (γ) activation of their contractile ends without any change of joint position. It seemed that such discharges could not signal joint position or movement unambiguously. Furthermore, intramuscular receptors were thought not to project to the cerebral cortex.

The demonstration of a cortical projection of intramuscular receptors set the scene for a reassessment of these views. Then, a variety of experiments on normal human subjects showed that proprioceptive signals based on intramuscular receptors can be perceived. These experiments included the following: (1) Anesthetization of the joints and skin of a finger or thumb, or of the whole hand, does not abolish a subject's ability to detect flexion or extension movements imposed on a relaxed digit. The ability must therefore be ascribed to discharges of sensory receptors in unaffected muscles in the forearm. (2) A similar but simpler experiment is to pull upon an exposed tendon in a conscious subject so as to stretch the muscle; this can be done while immobilizing the joint which the muscle normally moves. When this has been done most investigators have found that the subject reports that the joint that is usually operated by the pulled muscle seems to move, and to move in the direction that would normally pull the muscle. (3) When a human muscle is vibrated through the skin at about 100 Hz, an illusory joint rotation is experienced. The joint is felt to be moving in the direction that would usually stretch the vibrated muscle. This provides strong evidence that muscle spindles are responsible for the illusory sensations since they are known from electrophysiological experiments to be especially sensitive to such vibration. Vibration applied over the joint, but not over the muscle, does not cause the illusions, so joint receptors are not responsible.

The indication of a specific kinesthetic role for muscle spindles requires a solution to the problem of the potential ambiguity contained in their discharges because of their sensitivity to both stretch and fusimotor activity. A possible answer would be that the CNS informs itself internally of the level of fusimotor drive in any given circumstance, and that only the spindle discharge that exceeds what would be appropriate to that degree of fusimotor drive at a fixed muscle length has kinesthetic significance. Such a simple computation could be easily achieved by the CNS; essentially it amounts to nothing more than an appropriately scaled subtraction. Some evidence has been advanced to suggest that it is, in fact, done this way.

As the importance of intramuscular receptors for proprioception became apparent, there was a reassessment of the role of joint receptors. It was shown that little or no loss of proprioceptive acuity occurred when a joint, its capsule, and presumably its complement of joint receptors were removed surgically and replaced by an artificial joint. This, of course, need not imply that joint receptors do not contribute to proprioception but only that, if they do, their contribution can be duplicated from other sources, such as intramuscular receptors. An additional difficulty arose, however, when electrophysiological studies of receptors in some joints revealed few receptors which discharged in the mid-range of joint excursion. Moreover, the fibers that did discharge in this range were sometimes found not to alter their activity within the range. Thus, the question arose as to whether it might be the joint receptors that are unsuitable for kinesthesia. Further electrophysiological studies then found a larger proportion of mid-range receptors, many of them with well-modulated activity through the mid-range of joint rotation. Thus, it seems that at least for some joints, the electrophysiological behavior of the joint sensory nerves is suitable for contributing to proprioceptive sensibility.

A possible role for cutaneous receptors in kinesthesia has not, until recently, been seriously proposed or tested. Slowly adapting stretch receptors in the skin overlying joints might well provide information on joint position and movement. However, anesthetization of a sleeve of skin around the knee joint does not diminish proprioceptive acuity at that joint, whether or not it is combined with presumed anesthetization of joint receptors by intraarticular injection of local anesthetic. In the fingers and thumb, however, local anesthetization of both skin and joints seriously blunts kinesthetic sensibility, although surgical replacement of a digital joint (which leaves cutaneous sensation intact) has little effect. Therefore, a role for cutaneous receptors cannot be ruled out at least for the fingers. Possibly, any cutaneous contribution is one of facilitation of the central action of intramuscular and joint receptors. Indeed, one feature emerging from recent anatomical studies on proprioceptive pathways is the degree of convergence between cutaneous, joint, and intramuscular receptors at higher relay stations.

Proprioceptive acuity

Experimental work dating back to the 1890s established that proprioceptive acuity, measured as the ability to detect given angular displacements at given angular velocities, is superior in proximal joints to more distal joints (see Fig. 1). While this might seem to indicate an unusual concern of the CNS with proximal rather than distal body parts, it holds true only for data expressed in terms of angular rotations and velocities. When detections are expressed in terms of linear displacements and velocities of the terminal point of the limb moved by a particular joint (i.e., the finger tip in the case of the shoulder, elbow, and finger joints), proprioceptive acuity is ranked in the reverse order, i.e., finger best, shoulder worst. The mechanical arrangement of the joints explains this paradox: a given angular displacement at a proximal joint gives rise to a large displacement of the tip of the moved member while a similar angular displacement at a distal joint gives rise to a much smaller displacement. It is of interest that some proximal and distal joints have been demonstrated to give similar proprioceptive acuities when performance is expressed in terms of proportional changes and rates of change in the lengths of muscles operating at those joints, rather than in angular or linear terms.

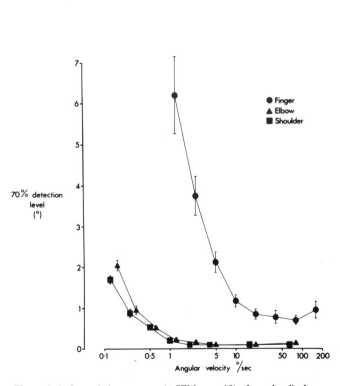

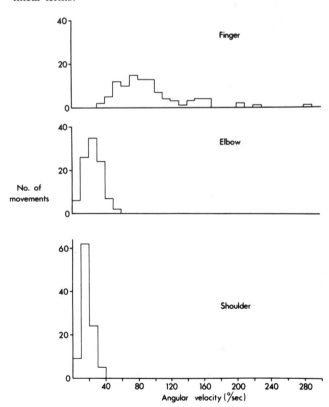

Figure 1. Left panel shows means (±SEM, $n = 10$) of angular displacements necessary for 70% correct detection (including correct nomination of direction) of randomly mixed flexions and extensions imposed at various angular velocities, on finger, elbow, and shoulder joints. Right panel shows distributions of angular velocities chosen by subjects in making accurate pointing movements using each of these joints. The angular velocities of voluntary movements correspond to the region of maximal proprioceptive acuity for each joint. Reproduced, with permission, from Hall and McCloskey (1983).

Voluntarily executed pointing movements requiring rotation of a given joint are carried out at velocities which correspond to the range of optimal proprioceptive acuity (see Fig. 1).

Force, tension, or heaviness

For conscious estimation of muscular forces or tensions as, for example, in judging the magnitude of an isometric contraction or even the weight of a lifted object, a variety of sensory receptors give useful signals related to the intramuscular tensions and to pressures on the skin in such circumstances. It has been shown, for example, that normal subjects are able to estimate and adjust intramuscular tension when only intramuscular receptors are available to provide guiding signals. Most normal subjects, however, prefer to disregard such signals from peripheral sensory receptors when judging forces or tensions and to rely instead upon estimating the magnitude of the centrally generated motor command, or effort, they employ to achieve a given muscular contraction.

The sensation of increasing heaviness of a suitcase that we carry with progressively fatiguing muscles is a common experience. Ultimately such a load is put down when it has ''become too heavy.'' But the load has not really become heavier: the pressures and tensions in the supporting limbs have not increased, and there is no reason to assume that the discharges from sensory receptors signalling pressures or tensions will have increased either. What makes the load seem heavier is that one perceives the greater effort, the greater efferent barrage of voluntarily generated command signals, that has been necessary to maintain a contraction with progressively fatiguing and so less responsive muscles. Similar sensations of heaviness or of increased muscular force accompany all other states of muscular weakness, whether caused experimentally (for example, by partial paralysis of muscles with curare) or by disease (perhaps most familiarly, as a result of simple stroke). It is the centrally generated motor command that signals muscular force or effort (albeit, in a nonlinear way)—an internally generated signal that is preferred to the conventional and often more correct signals arising in sensory receptors. However, the centrally generated signal gives no sensation of movement or of altered position. Thus, for example, a phantom limb feels heavy, but elements within the phantom are not perceived to move in relation to each other in response to an attempt to move.

One cannot judge the weight of an object if a muscular effort cannot move or support it. This must be so whether judgment about weights is usually based on the discharges of peripheral sensory receptors or centrally generated, command-related discharges. As one attempts to lift a heavy object, centrally generated commands increase through the range of the effort, and peripheral sensory signals related to the effort also vary through this range. Unless one knows which of the range of commands or sensory signals is associated with lifting or supporting the object, one cannot say how heavy an object is—only that it is heavier than some other object that can be lifted or supported. It has been shown that for estimating weights it is sufficient for the nervous system to be provided with a quite crude signal, usually a proprioceptive signal that the body part exerting the force has managed to move, to indicate which command in a range of motor commands is sufficient to move the weight.

Perceived timing of movements

As there are sensory inputs from muscles, joints, and skin available to signal positions and movement, and a sense of effort or muscular force as well as these for signalling force or heaviness, there are clearly many sources of information on which one might rely when judging the timing of various muscular contractions. In studies on the perceived timing of muscular contractions, it has been found that most normal subjects can reliably define a moment prior to the appearance of electromyographic activity in their target muscles as coincident with their dispatch of a command to move. They can also reliably define a moment well after the appearance of electromyographic activity as coincident with the onset of actual movement. Thus, it appears that appropriate testing reveals two separate moments that are subjectively associated with the onset of movement. It seems likely that centrally generated, internal neural signals give rise to the first percept, and afferent proprioceptive signals to the second. Apparently, in normal movements, the two percepts are usually fused, or one is disregarded.

Further reading

Burgess PR, Wei JY, Clark FJ, Simon J (1982): *Ann Rev Neurosci* 5:171–188

Hall LA, McCloskey DI (1983): Detections of movements imposed on finger, elbow and shoulder joints. *J Physiol (Lond)* 335:519–533

McCloskey DI (1978): Kinesthetic sensibility. *Physiol Rev* 58:763–820

Lateral Line System

Peter Görner

The lateral line system consists of water-movement-sensitive sensory organs in the epidermis of cyclostomes, fishes (Chondrichthyes, Osteichthyes), and aquatic stages of all three orders of amphibians. The phylogenetic origin of this system is not known. Already the oldest known ancestors of the modern fishes, the ostracoderms, probably possessed a lateral line system. At least twice during evolution part of this system evolved to an electrosensitive system, possibly by reduction of mechanical sensitivity and enhancement of electrical sensitivity of the sense organs. Like the labyrinth, the lateral line system derives from dorsolateral placodes.

Morphology

In cartilaginous and bony fishes parts of the sense organs are submerged in subepidermal canals or grooves on the head and trunk (lateral line). A lateral line organ consists of one

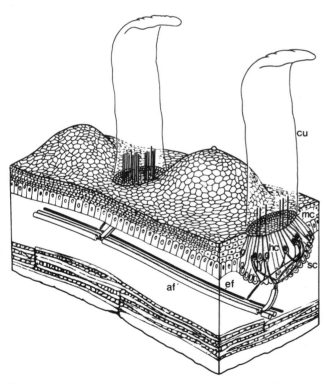

Figure 1. Semischematic view of two neuromasts in a lateral line stitch of *Xenopus laevis*. af, afferent fibers; cu, cupula; ef, efferent fibers; hc, hair cells; mc, mantle cells; sc, supporting cells.

neuromast (canal organs, some epidermal organs) or several neuromasts (most epidermal organs), arranged singly or in groups or rows (stitches, Fig. 1). Epidermal neuromasts may also lie in slits or pits (pit organs in elasmobranchs or auricles in catfish). On the dorsal and ventral surface of the rostral part of electric rays, three neuromasts (one central, two lateral) form a Savi vesicle, 100 to 200 of which are arranged in a row. The typical budlike neuromast is composed of elongated mantle cells enveloping a core of supporting cells and several shorter cylindrical or flask-shaped sensory cells (hair cells). On the apex of the hair cell one long stiff cilium (kinocilium) with a typical 9 + 2 arrangement of the filaments is inserted excentrically, and about 50 considerably shorter stiff stereovilli (stereocilia) arise and decrease in length with increasing distance from the kinocilium. There are two sets of hair cells with respect to the insertion of the kinocilium. In approximately half the hair cells, the cilium, inserting near the rim of the cell apex, is positioned toward one end of the oval sensory epithelium of a neuromast; in the other half of the hair cells the cilium is positioned toward the opposite end. The neuromast is covered by a jelly-like cylindrical or more or less flattened cupula. The cupula is composed of two layers, a central core secreted by the supporting cells, and a surrounding sheath secreted by the mantle cells. The cupula continuously grows proximally (15 to 30 μm per day in *Necturus*) and erodes distally. It is nearly critically damped and most probably determines the mechanical frequency characteristic of the hair cells within the neuromast.

A neuromast is innervated by at least two (urodels, trunk organs of anurans) afferent and, with the exception of cyclostomes, one or more myelinated efferent nerve fibers. It is generally accepted that one afferent fiber innervates only one of the two sets of hair cells, i.e., hair cells with the same alignment of the kinocilium, while efferent fibers innervate both sets. The diameters of the myelinated fibers range from 1 to 15 μm in teleosts and amphibians and 2 to 25 μm in elasmobranchs. The efferent fibers synapse on the hair cells. The thicker fibers are always afferents, and the thinner fibers may be efferents or afferents (teleosts, elasmobranchs). Additionally, the neuromasts are innervated by a few unmyelinated efferent fibers with diameters below 1 μm. In the clawed toad *Xenopus* they have been shown to be of sympathetic origin.

The lateral line afferents enter the medulla rostrally (anterior lateral line) or caudally (posterior lateral line) of the nucleus octavus and bifurcate into an ascending and a descending tract. Collaterals of the descending tract terminate in the medullary nucleus medialis (n.m.) and nucleus caudalis (n.c.). Ascending fibers end in the rostral part of the n.m. and, in teleosts and amphibians, in the eminentia granularis of the cerebellum. In teleosts they may also project into the corpus cerebelli and into the valvula cerebelli. No indication for a somatctopic organization has been found in the termination pattern of the primary afferents or in the arrangement of secondary neurons within the medullary lateral line nuclei (n.m. and n.c.). Secondary neurons of the n.m. and n.c. project to the contralateral n.m. and n.c. and probably establish an inhibitory pathway between the left and right side. The other secondary neurons project to the contralateral and, to a small amount, to the ipsilateral torus semicircularis (t.s.) in the midbrain. In contrast to the medullary nuclei there is evidence for a somatotopic organization in the t.s. A cluster of neuromasts projects point to point to a cluster of central neurons in the t.s. The fact that besides lateral line activity, activity of other sensory systems (labyrinth organs, somatosensory organs) may converge on the same neuron, can be quoted as an argument for the prominent integrative function of the t.s.

The efferent somata (a few in amphibians and 10 to 40 in teleosts) form two groups in the formatio reticularis of the medulla. In teleosts additionally a few cells are found in the diencephalon near the zona limitans. Several of the efferent cells in the formatio reticularis also innervate hair cells of the octavus system.

Electrophysiology

The resting potential of different hair cells may vary between 10 and 65 mV in the burbot (*Lota lota*). A hair cell has a directional sensitivity following a cosine wave function. It is depolarized when the cilium is bent ("sheared") away from the stereovilli and is hyperpolarized when it is bent in the opposite direction. On repeated stimulation with 27 Hz the receptor potential is maximally 800 μV peak to peak in *Necturus maculosus*. Most afferent nerve fibers are spontaneously active with an impulse rate from below 1 Hz (elasmobranchs, teleosts) to about 100 Hz (teleosts) and a rarely regular, mostly irregular Poisson or gaussian distribution of the impulse intervals. When the hair cells of a neuromast are depolarized by a constant bending of the cilia, the impulse rate first increases and then declines to a constant value (phasic-tonic reaction). Superficial neuromasts react to water displacement from ca. 0.1 μm on and are most sensitive to a stimulation frequency between 20 and 40 Hz in amphibians and 10 and 50 Hz in teleosts. Canal neuromasts of teleosts are even more sensitive. They respond already to 0.01 μm water displacement (perpendicular to the canal pore) within the most sensitive frequency range of 70–180 Hz. The slope of the frequency transfer function lies between 6 and 12 dB/octave, indicating that the lateral line activity is correlated either with water velocity or with water acceleration. Secondary neurons in the lateral line system have large receptive fields and show low-pass filter characteristics: They do not follow stimulus intervals below 10 msec with a one-to-one reaction when the primary lateral line fibers are stimulated electrically.

The significance of the efferent system is not yet fully understood. Stimulation of the efferent fibers normally results in a reduction of the afferent lateral line activity. The efferents are activated (5–10 Imp/sec in the dogfish) primary to and during movement and also when other sensory organs are stimulated. Stimulation of the lateral line system itself may lead to an activation of the efferent fibers. The maximal attenuation of afferent activity during swimming reaches about 15 dB. The transmitter at the efferent synapse is most probably acetylcholine. Besides the hair cells, second-order afferent neurons in the medullary n.m. and n.c. of bony fishes are inhibited due to efferent input. The hyperpolarization in these neurons during active movement is 4–6 mV. By this, adaptation is prevented and the sensitivities of primary and second-order sensory neurons are preserved.

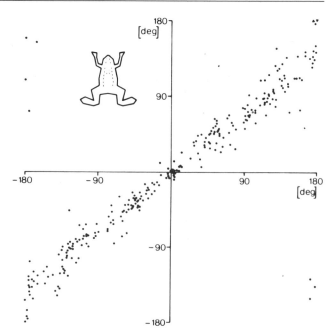

Figure 2. Turning angles (ordinate) of three *Xenopus laevis* to surface waves that were elicited by dipping a rod into the water at a distance of 12 cm. Abscissa: stimulus angles.

Behavior

Compared with our knowledge about the organization of lateral line organs little is known about the biological significance of the lateral line system. Enucleated bony fishes and amphibians are able to localize a scrambling prey object or a moving or oscillating dummy in their vicinity (*Ferntastsinn*, i.e., touch at a distance). Clawed toads turn toward the center of a surface wave by means of lateral line input with remarkable accuracy (Fig. 2). Surface feeding fishes of several orders that have developed special morphological adaptations for the perception of surface waves are able to localize the wave center accurately up to a distance of 3 to 3.5 body lengths. Thereby the relevant information is the curvature and the frequency spectrum of the waves passing the animal as well as the frequency-to-amplitude ratio of the first 8 to 10 cycles within the wave train. The blind cave fish (*Anoptichthys jordani*) orientates itself by means of the lateral line system while gliding. It can detect bars of 2 mm diameter when passing at a distance of a few millimeters, and discriminate between two fences of 6 and 4 bars. The lateral line system is important in schooling of saithe (*Pollachius virens*) and most probably also in schooling of *Xenopus* tadpoles.

Further reading

Bleckmann H (1985): The role of the lateral line. In: *The Behaviour of Teleost Fishes*, Pitcher J, ed. Sydney: Croom Helm Australia

Dijkgraaf S (1962): The functioning and significance of the lateral-line organs. *Biol Rev* 38:51–105

Russell IJ (1976): Amphibian lateral line receptors. In: *Frog Neurobiology*, Llinas R, Precht W, eds: Berlin Heidelberg New York: Springer-Verlag

Sand O (1984): Lateral line systems. In: *Comparative Physiology of Sensory Systems*, Bolish L, Keynes R, Madrell SHP, eds. Cambridge London New York Melbourne: Cambridge University Press

Mechanoreceptors

Paul R. Burgess, Kenneth W. Horch, and Robert P. Tuckett

Mechanoreceptors are widely distributed throughout the body, being found in skin, muscles, articular tissues, visceral and thoracic organs, connective tissues, and blood vessels. Wherever mechanoreceptors are found, two questions arise: How are the receptors to be classified? And what sorts of information do they provide about mechanical stimuli? We will consider cutaneous mechanoreceptors as an example of how these issues might be addressed. The general principles derived from this population of receptors also apply to mechanoreceptors in deeper tissues.

The varied experiences evoked by mechanical stimulation of the skin range from faint, evanescent tickles to strong and persistent sensations of indentation or pressure. To simplify matters, consider a single stimulator that indents the skin at some rate, remains more or less stationary for a time, and then retracts to the starting position. Let us assume that the portion of the stimulator that contacts the skin (the stimulator face) is smooth, otherwise there might be more than one locus of cutaneous stimulation. Under these conditions we can subdivide the parameters defining the stimulus into three categories: (1) those that act in a direction orthogonal to the plane of the skin; (2) those that specify events occurring parallel to the receptor sheet; and (3) parameters concerned with how events affecting the first two components are distributed in time.

Parameters in the orthogonal dimension include the depth of indentation or pressure at any moment and the direction (advancing into the skin or retracting) and rate of change of indentation or pressure. The pressure (force per unit area) with which a stimulator pushes against the skin is determined by the location on the skin being stimulated, the depth of indentation, the time for which the indentation has been maintained, and the direction and rate of change of the indentation. This complex function is difficult to predict analytically but has an important influence on the perceived strength (intensity) of a stimulus. Parameters in the surface plane include the location of the stimulator on the body surface, the size and configuration of the stimulator face, which determines the area and shape of the patch of skin being indented, and the contour of the indented area as determined by the curvature of the stimulator face. Temporal parameters specify the duration and pattern of any fluctuations in the stimulus.

The question of what kinds of information cutaneous mechanoreceptors convey about these parameters is intimately tied to the question of how the receptors are to be classified. Both of the two general methods of classifying cutaneous receptors, morphological and physiological, provide insight into the functional role of these sense organs.

Most cutaneous mechanoreceptors can be classified as belonging to one of the eleven groups listed in Table 1, which also shows morphological and response properties for each group. Receptors responding to maintained indentation or displacement of the skin or a hair are called tonic; those requiring movement of the innervated structure to elicit a response are

Table 1. Types and Properties of Cutaneous Mechanoreceptors

Type	Axon Size	Adequate Stimulus	Receptor Structure
PC	A-alpha	acceleration of skin and deeper tissue	Pacinian corpuscles
G1 hair	A-alpha	acceleration or velocity of hair movement	endings around hair follicles
GI hair	A-alpha	velocity of hair movement	
G2 hair	A-alpha	displacement and rate of displacement of hair from rest position	
F1 field	A-alpha	velocity of skin indentation	Meissner's corpuscles and related sparsely laminated capsular endings
FI field	A-alpha	velocity of skin indentation[a]	
F2 field	A-alpha	indentation and rate of indentation of skin	
T1	A-alpha	indentation of touch disk[b]	Merkel cell complex
T2	A-alpha	skin stretch[b]	Ruffini ending
D mechano	A-delta	velocity of movement of hair or skin	free nerve endings?
C mechano	C	lingering indentation or slow rates of indentation of skin	free nerve endings?

[a] Intermediate field receptors respond to lower rates of indentation and show more directional sensitivity than F1 receptors.
[b] The responses of Type I and Type II receptors are also influenced by the rate of touch disk indentation or rate of skin stretch, but both produce a continuous discharge to a maintained displacement. F2 receptors will cease to respond to a maintained indentation within 30 to 40 sec.

called phasic. In general, the conduction velocity of fibers innervating phasic mechanoreceptors is faster than the conduction velocity of afferent fibers innervating tonic receptors. For nerve fibers innervating receptors on the hindlimb, those serving phasic receptors are more likely to have a branch that ascends directly to the medulla than are those associated with more tonic mechanoreceptors.

The more tonic mechanoreceptors respond to both the displacement of the innervated structure from the rest position and the rate of change of displacement (velocity). Tonic receptors are better activated by movements away from the rest position than by movements back toward the rest position. That is, they are directionally sensitive. The most phasic mechanoreceptors, in contrast, are not directionally sensitive: they respond nearly equally well to movements toward the rest position as to movements away from the rest position.

In addition to the type of movement in which the skin or hair is displaced by some amount from its rest position, held at this new position for some time, and then returned to the starting position, one can consider how cutaneous mechanoreceptors respond to vibratory stimuli. On theoretical grounds, one would expect that the more phasic receptors would respond better to high frequencies than would the more tonic receptors. This is what is found empirically. If one measures the sensitivity of the receptors by looking at the minimum amplitude displacement required to elicit some criterion threshold response as a function of sinusoidal frequency, one finds that the slope of these tuning curves is steeper for phasic receptors than it is for tonic mechanoreceptors. Intermediate cutaneous mechanoreceptors fall in between these two extremes in threshold level and tuning curve slope.

It is clear from these considerations that the response properties of the array of mechanoreceptors found in the skin is sufficiently complex to provide for the rich and varied repertoire of sensations elicited by cutaneous stimulation. We have yet to learn to what extent somatic sensation depends on synergistic interactions between these receptor types or to what degree different receptor types contribute distinct types of information. Although the classification used here makes use primarily of a receptor's ability to signal the speed, depth, and direction of a stimulus, individual receptors generally do not provide information about any one of these parameters with high purity. Central circuits process and shape the information from the receptors to create the surprisingly reliable information we have about the external world.

Further reading

Darian-Smith I (1984): The sense of touch: Performance and peripheral neural processes. In: *Handbook of Physiology. Section I: The Nervous System. V. III. Sensory Processes, Part 2*, Darian-Smith I, ed. Bethesda: American Physiological Society

Gordon G, ed (1977): Somatic and visceral sensory mechanisms. *Br Med Bull* 33:89–177

Horch KW, Tuckett RP, Burgess PR (1977): A key to the classification of cutaneous mechanoreceptors. *J Invest Dermatol* 69:75–82

Hunt CC, ed (1974): *Handbook of Sensory Physiology. Muscle Receptors. V. 3/2*. New York: Springer-Verlag

Motion Sense

Volker Henn

The sense of self-motion can be described as the awareness of motion of our head and body, be it active, as in walking, or passive, as in moving in a vehicle. It also gives us information about our orientation relative to gravity. This sense is unique among all the senses in that it does not depend on a single sensory receptor organ. Its first scientific description dates back to Ernst Mach with his publication in the year 1875 of *Grundlinien der Lehre von den Bewegungsempfindungen* (*Fundamentals of the Theory of Motion Perception*). In humans, motion sense primarily utilizes information from the labyrinths and from the visual and somesthetic systems. Other sensory organs such as the acoustic system may play a greater role in some animals (e.g., owl). The labyrinths in the inner ear detect linear and angular acceleration. This information is transformed into electrical activity by the hair cells which in turn modulate activity in the vestibular nerve. The nerve projects to the vestibular nuclei in the brain stem. At this level the labyrinthine input converges with motion-specific information from other sensory systems. For example, imagine sitting on a turntable and being rotated in the light. In this situation you can give a very accurate estimation of the subjective velocity of rotation or change in velocity (acceleration or deceleration). Imagine the same situation in darkness. You now only sense accelerations, i.e., changes in velocity but not rotation at a constant velocity. If rotation at a constant velocity continues, the rotatory sensation attenuates and completely disappears within seconds, only to reoccur when one is stopped, i.e., during negative acceleration. This information about acceleration is conveyed by the labyrinths. The different sensation that occurs during the same rotation in the light is due to the visual contribution to motion sense. It can be tested separately by rotating the visual surround around a stationary subject. Experimentally this is usually done by placing the subject inside a large drum which can be rotated at controlled speeds. If held stationary inside an illuminated drum that slowly rotates, one invariably experiences the drum as stationary and oneself as rotating. This very compelling sensation of a visually induced sensation of self-rotation is called circularvection. It is so compelling because the visual information directly activates the vestibular nuclei, inferred from animal experiments. Activation of the vestibular nuclei with their projection to the thalamus and cortex leads to the sensation of motion, independent from which sensory system these nuclei were activated.

In an analogous manner, the otoliths, which detect linear acceleration, project centrally to the vestibular nuclei. There this information interacts with inputs from other sensory organs, mainly the visual and somesthetic systems.

Because of this multisensory convergence, the loss of labyrinthine function does not lead to the kind of complete deficit that would be comparable to blindness with the loss of the eyes or deafness with the loss of inner ear function. As there is a multiple cortical presentation with input from the vestibular and other sensory systems in motion sensation, loss of self-motion perception has not been reported as a neurological symptom. This may well explain why a motion sense is not categorized as one of the five classical senses of smell, vision, taste, hearing, and touch.

If a normally present input to the vestibular nuclei is absent, one experiences vertigo. An example is space flight where the linear acceleration vector of gravity is missing. It takes several days to adapt to such a situation of changed sensory inputs. In a similar way, if one of the inputs is affected by pathology, the initial sensation of vertigo usually attenuates even if the normal function has not been restored.

This use of information from many different receptor organs makes the sense of motion unique. All other sensory organs derive their specificity from the process of sensory transformation at their respective sensory cells, like the light sensitivity in the retina and the mechanical sensitivity of skin receptors. The specificity of the sense of motion relies on the fact that from the wealth of information gathered by the different sensory organs just one aspect, that of motion, is specifically conveyed to central vestibular structures giving rise to the subjective sensation of motion as well as leading to motor reflex responses like nystagmus and posture stabilization.

Further reading

Henn V (1984): E. Mach on the analysis of motion sensation. *Human Neurobiol* 3:145–148

Henn V, Cohen B, Young LR (1980): Visual-vestibular interaction in motion perception and the generation of nystagmus. *Neurosci Res Prog Bull* 18:457–651

Motion Sickness

Ashton Graybiel

The term "motion sickness" was proposed by Irwin in 1881 to provide a general designation for such similar syndromes as seasickness, train sickness, and the like. This term, imprecise for scientific purposes, has gained wide acceptance because it meets the test of convenience by its etiologic and symptomatic connotations.

Motion sickness has never been defined to everyone's satisfaction, hence the resort to description. Frank motion sickness comprises a constellation of symptoms resulting from too sudden an exposure to periodic unnatural accelerations, and its course is self-limited through the mechanisms of adaptation. Loss of function of the vestibular organs confers immunity, indicating their essentiality in the genesis of the symptoms, but secondary etiologic factors tending either to increase or decrease susceptibility are always present.

Normal persons are inherently susceptible in greater or less degree, and susceptibility in the same person may differ in different gravitoinertial force environments. Overt manifestations (Table 1) include pallor, sweating, salivation, drowsiness, and, most important from a practical standpoint, activation of the vomiting reflex centers. First-order responses give rise to second-order responses and complications of ever increasing diversity that may lead to prostration. The diagnosis of frank motion sickness is so easy that it is not a matter of professional concern, but mild symptoms may pose a diagnostic problem.

Prevention is the key to good management in all cases. Motion sickness, theoretically, is preventable, but countermeasures may be impractical under many circumstances. Once symptoms are well established, quick relief may require intensive therapy.

Despite the prevalence of motion sickness and its challenge to the investigator, little was accomplished under field conditions toward elucidating its underlying mechanisms. Even under controlled laboratory conditions, progress has been slow.

Vestibular systems

The nonacoustic portion of the inner ear (Fig. 1) comprises the otolith apparatus and semicircular canals, collectively termed the vestibular organs. In each ear the sensory receptor mechanisms in the three mutually perpendicular canals are stimulated by angular and Coriolis accelerations, generated normally by rotary motions of the head and body. The paired otolith organs, their receptor mechanisms in the shape of four curved plates constituting, in toto, a significant portion of a sphere, are stimulated continually by gravity and also by impulse linear accelerations. The vestibular organs are encased in a bony labyrinth, and, while protected from external mechanical stimuli, must react slavishly to accelerative forces insofar as their response characteristics permit. The important point to be made here is that exposure to motions in convey-

Table 1. Diagnostic Categorization of Different Levels of Severity of Acute Motion Sickness

Category	Pathognomonic 16 Points	Major 8 Points	Minor 4 Points	Minimal 2 Points	AQS[a] 1 Point
Nausea syndrome	nausea III,[b] retching or vomiting	nausea II	nausea I	epigastric discomfort	epigastric awareness
Skin		pallor III	pallor II	pallor I	flushing/subjective warmth ≥ II
Cold sweating		III	II	I	
Increased salivation		III	II	I	
Drowsiness		III	II	I	
Pain					headache (persistent) ≥ II
Central nervous system					dizziness (persistent) eyes closed ≥ II eyes open ≥ III
Levels of severity identified by total points scored					
	frank sickness (FS) ≥ 16 points	severe malaise (M III) 8–15 points	moderate malaise A (M IIA) 5–7 points	moderate malaise B (M IIB) 3–4 points	slight malaise (M I) 1–2 points

[a] AQS, additional qualifying symptoms.

[b] III, severe or marked; II, moderate; I, slight.

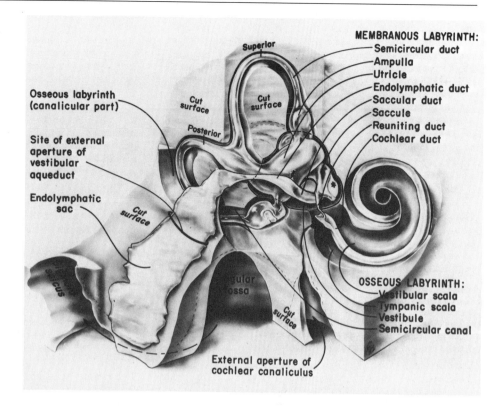

Figure 1. Reconstruction of the membranous labyrinth and related anatomy. From Anson, B. G., Harper, D. B., and Winch, T. G. (1968): Third Symposium on the Role of the Vestibular Organs in Space Exploration. NASA SP-152.

ances or other devices may stimulate the vestibular organs by virtue of (1) angular, linear, and Coriolis accelerations generated by two angular or linear velocities, one contributed by humans and one by machine, and (2) the accelerative force generated by passive rotation of a subject at constant velocity in a device tilted away from the vertical.

The vestibular organs have a large representation in the brain stem, cerebellum, and spinal cord. By virtue of an incredibly extensive and complex neuronal circuitry, mainly at subcortical levels, the vestibular servation systems participate in such functions as the maintenance of postural equilibrium and the coordination of eye-head-body motions. Evolutionary adaptation has ensured that such participation is effected in elegant fashion in naturally occurring accelerative environments, but this does not hold true for exposure to all types of artificially generated accelerations. This limitation accounts for the great individual and intraindividual differences in susceptibility to motion sickness but, of course, explains nothing.

Mechanisms and manifestations

When a susceptible person is exposed to an adequate accelerative stressor, the resulting abnormal pattern of afferent vestibular impulses exceeds the capacity of the vestibular servation systems to maintain homeostasis with two distinguishable consequences. The first is confined to the vestibular systems proper and is declared by perturbations of normal responses. The second involves a failure in homeostasis allowing the irradiation of vestibular activity to reach extraneous servation systems, and the manifestations evoked constitute an epiphenomenon superimposed on any manifestation of the first and includes motion sickness. The manifestations having their immediate origin in nonvestibular systems far exceed the overt signs and symptoms of motion sickness and are absurd in terms of the needs of the organism. Recently, there has been

a systematic attempt to categorize these manifestations as first-, second-, or third-order effects and their complications. The release of antidiuretic hormone and appearance of stress hormones (catechols and corticoids) in the urine testify to the general nature of the symptomatology when the stress is sufficient and sufficiently prolonged.

The recovery from motion sickness during continual exposure to the stressful accelerations is complicated. First, the nonvestibular servation systems must be freed from outside influences (vestibular activity, chemical agents), after which restoration takes place spontaneously through homeostatic events and processes. The time course of these events overlaps and has not been clearly defined. Curves depicting responses evoked on sudden return to a stationary environment depend on the level of adaptation acquired in the vestibular system and whether restoration was complete in nonvestibular systems. The *"debarquement"* has been simulated in the laboratory. If adaptation is incomplete, both ataxia and motion sickness are experienced; if complete, manifestations are limited to ataxia and dizziness.

Persons vary greatly in their susceptibility to motion sickness, a small percentage manifesting either high or low susceptibility and the great majority varying within the average range. The decreasing susceptibility with increasing age has been explained by adaptation. Most persons retain some of the adaptation acquired, and adaptation effects acquired in one situation transfer, in part, to other situations.

Drowsiness is one of the cardinal symptoms of motion sickness; therefore, a symptom complex centering around drowsiness has been identified which, for convenience, has been termed the sopite syndrome. Generally, the symptoms characterizing this syndrome are interwoven with other symptoms of motion sickness, but under two circumstances the sopite syndrome comprises the main or sole overt manifestation of motion sickness. One circumstance is that in which the intensity of the eliciting stimuli is closely matched to a person's susceptibility, and the sopite syndrome is evoked either before other symptoms of motion sickness appear or in their absence. The second circumstance occurs during prolonged exposure in a motion environment when adaptation results in the disappearance of motion sickness symptoms, except for responses characterizing the sopite syndrome. Typical symptoms of the syndrome are: (1) yawning; (2) drowsiness; (3) disinclination for work, either physical or mental; and (4) lack of participation in group activities. Phenomena derived from an analysis of the symptomatology of the sopite syndrome are qualitatively similar but may differ quantitatively from abstractions derived in other motion sickness responses.

One example is the unique time course of the sopite syndrome. This implies that the immediate eliciting mechanisms not only differ from those involved in evoking other symptoms but also that they must represent first-order responses. Diagnosis is difficult unless the syndrome under discussion is kept in mind. Prevention poses a greater problem than treatment.

Anti-motion-sickness remedies

Drugs with central anticholinergic actions such as scopolamine, promethazine, dimenhydrinate (Dramamine), cyclizine (Marezine), and meclizine (Bonine) are effective in increasing resistance to motion sickness. Drugs with central effects that enhance the action of norepinephrine are also effective. These include dextro-amphetamine sulfate (Dexedrine) and ephedrine. It is recommended that all anti-motion-sickness drugs be taken at least one hour before exposure to motion. For severe conditions of motion it is recommended that one of the following drug combinations be used: (1) scopolamine, 0.6 mg, plus dextro-amphetamine sulfate, 5 mg, or (2) promethazine, 25 mg, plus ephedrine, 25 mg. Moderate conditions may be controlled by dimenhydrinate, 50 mg, and in mild conditions such as automobile travel, cyclizine, 50 mg, or meclizine, 50 mg, may be sufficient.

Views presented are those of the author and do not necessarily reflect those of the Navy Department.

Further reading

Graybiel A, Knepton J (1976): Sopite syndrome: A sometimes sole manifestation of motion sickness. *Aviat Space Environ Med* 47:873–882

Graybiel A, Wood CD, Miller EF II, Cramer DB (1968): Diagnostic criteria for grading the severity of acute motion sickness. *Aerospace Med* 39:453–455

Irwin JA (1881): The pathology of sea-sickness. *Lancet* 2: 907–909

Wood CD, Graybiel A (1972): Theory of antimotion sickness drug mechanisms. *Aerospace Med* 43:249–252

Multisensory Convergence

Peter H. Hartline

Much of the research on multisensory or multimodal integration has been directed toward (1) learning what neurons or regions of the brain are specialized for responding to objects that can stimulate more than one sensory modality (multiple modality objects), (2) determining the functional properties of multisensory neurons and inferring their roles in behavior, and (3) determining how structural properties of such neurons may account for their modality-integrating function.

Classification of multisensory neurons

Among invertebrates and vertebrates, there are many examples of neurons that are excited or modulated by input from more than one sensory system. The most common type of multimodal neuron, termed OR neuron, responds reliably to stimuli that activate any of two or more senses (e.g., visual or infrared senses in pit vipers, visual or auditory modalities in mammals, visual (ocella) or wind hair in insects). The number of spikes evoked by simultaneous stimulation of two senses exhibits linear summation or intermodality occlusion (less than linear summation). Some OR neurons exhibit enhancement (more than linear summation) or depression (fewer spikes than would be evoked by the more effective modality alone). No studies demonstrate that neurons downstream from a particular multisensory neuron can identify the sensory origin of the message (e.g., by a candidate pattern code). Thus, there is no evidence that OR neurons are serving as multiplexed sensory channels.

Illustrative examples of other functional categories of cells (excluded from the OR category because they are not reliably excited by single-modality stimulation of each of their sensory inputs) follow: (1) Infrared ENHANCED visual neurons of rattlesnakes are driven by single modality visual but not infrared stimuli, but combining the infrared with the visual stimulus yields a substantially greater response. Similar modality combination rules hold for visual ENHANCED auditory or vibrissal neurons of mammalian optic tectum, etc. (2) Visual DEPRESSED vibrissal neurons of rodents are excited by touching an appropriate facial whisker but not by visual stimuli. Combined stimulation generates fewer nerve impulses than vibrissal stimulation alone. (3) Infrared-visual AND neurons of rattlesnakes do not reliably respond to either single-modality stimulus, but respond well to combined stimulation. Interactions such as these have been found for diverse combinations of modalities in many species; they probably represent multisensory integration mechanisms common to vertebrates.

Functional roles of multisensory neurons

OR neurons would be appropriate for a *sensory substitution* function—one in which input via any active modality can provide the information needed for downstream processing. For example, locust ocellar and wind hair sensor inputs converge on interneurons that help stabilize the insect's flight. In many vertebrate species, OR neurons in the optic tectum (superior colliculus in mammals) probably allow an object to trigger an orientation movement via several modalities. An example of higher order cells that manifest OR type sensory integration is provided by hypothalamic neurons, recorded in primates, that respond to sight or taste of food.

Since actions initiated in response to sensory information must be carried out by a common set of effectors, one is not surprised to find multiple senses represented in the activity of elements in the *final common path* of motor control. An interneuron that triggers a locust's jump, for instance, can be excited by tactile, auditory, or visual stimuli. In fish, the Mauthner neuron (which triggers the tail flip) can be fired by input of lateral line, visual, or other modalities. In these and other examples, elements of the final common path may have nontrivial decision-making functions (as in the Mauthner cell's choice of a left or right flip of the tail, which must reflect laterality of the composite sensory stimuli).

OR neurons might also serve a *noise reduction* function. For instance, some insect OR neurons follow the temporal pattern of an artificial song more faithfully when the song is presented simultaneously to auditory and substrate vibration receptors than when it is presented to auditory receptors alone.

The intermodality facilitation of ENHANCED and AND neurons could be a basis for several candidate functions: (1) *multimodal recognition* (e.g., snake infrared-visual AND neurons might participate in recognition of edible warm-blooded prey), (2) *cross-modality attention* (in which one sense primes appropriate neurons for vigorous response to a different sensory input), and (3) *noise reduction* (as in OR neurons). This would be advantageous for multimodal detection or localization of objects that generate near-threshold stimulation in more than one modality (ENHANCEMENT tends to be numerically more impressive as the stimuli approach threshold; furthermore, two subthreshold stimuli can interact to evoke a clear response).

The intermodality interactions of DEPRESSED neurons could participate in *multimodal recognition* by suppressing the response to objects having a particular multisensory signature. Depression might also occur as a by-product of other mechanisms, e.g., it might reflect inhibition obtained by stimulating an antagonistic surround of an ENHANCED cell. ENHANCED and DEPRESSED neurons, as well as OR neurons, have been found in the optic tectum, in thalamic nuclei, and in several regions of frontal, parietal, and temporal association cortex.

One might expect temporal simultaneity of multisensory stimulation to be a requisite for nonlinear interactions. This has not generally been borne out by studies of neurons in superior colliculus or association cortex of mammals; instead, maximal interaction can occur at a nonzero intermodality delay, and the same neuron may show ENHANCED or DEPRESSED interactions depending on the intermodality stimulus interval.

Whether this indicates a functional role of such neurons in detecting *sequence of sensory cues* is not known.

It is unclear to what extent complex features of a stimulus object are recognized by operationally similar feature-detecting mechanisms for each modality of a multisensory neuron. A few reports indicate that similar directional motion preferences for visual and nonvisual stimulation are expressed in some multimodal cells (e.g., the direction of motion preferred by some auditory-visual neurons of mammals and infrared-visual neurons of snakes is the same for both senses).

Multisensory representation of space

The most prominent and universal specialization for responding to multiple modality objects is the connectivity that (1) brings into alignment the spatiotopic maps of two senses that project to the same brain region, and (2) aligns spatial receptive fields of multisensory neurons. Such alignments doubtless reflect the fact that an object stimulating two modalities at once occupies, at any time, a single location in space. Examples are found in (1) the optic tectum, where convergence among visual, somatic, infrared, auditory, and electrosensory modalities occurs in diverse species from elasmobranchs through primates, and (2) the visual cortex, where visual, auditory, and probably tactile senses converge.

The optic tectum is believed to play an important role in initiating movements that orient an animal's eyes, head, or body toward a source of stimulation. Both its input (sensory) and output (to motor centers) reflect a common spatiotopic mapping principle. Along the rostrocaudal axis in most species, more caudal locations represent more temporal or posterior visual space (on the side contralateral to the tectum under consideration) and posteriorly or laterally directed action (e.g., movement of the eyes); along the mediolateral axis, more medial locations represent more dorsal visual space and upwardly directed action. Until recently, the tectum was regarded as a predominantly visual or visuomotor center. It is now known that most nonvisual senses that might be useful for spatial localization are also prominently represented in the tectum. The nonvisual senses follow a spatiotopic plan similar to that for vision. Thus, the rostral pole of pit viper's left tectum contains infrared-sensitive neurons with nasal (anterior) receptive fields on the snake's right, the rostral pole of the owl tectum contains auditory neurons whose best areas (most responsive part of the receptive field) are directed near to the frontal midline, and the rostral pole of the tectum of skates and weakly electric fish have neurons responsive to anteriorly located electrical perturbations (insulators, conductors, current sources). In primates, rapid adjustment of the nonvisual spatiotopic plan may compensate for voluntary eye movements.

Tactile senses can also convey spatial information and are represented in the optic tectum. Whereas the nonvisual senses described above report on objects that are remote from the animal's body, tactile senses report stimulation by objects that contact (or come very near to) the body. Not surprisingly, tactile inputs are also spatiotopically organized. More anterior vibrissae and other tactile sense organs of mammals, reptiles, amphibians, and fish are represented in the rostral part of the tectum; more dorsal areas of the body surface are represented more medially. This spatial representation scheme, because it is common to many senses (whether they deal with remote space or body surface), appears optimum to allow *sensory substitution,* in which localized sensory cues lead to the appropriately oriented movement regardless of the input modality.

In each of several distinct regions of visual cortex, visual inputs are organized according to a spatiotopic map, as in the optic tectum. The visual cortex is thought to participate in processes subserving recognition of objects or patterns. Though this part of the brain is considered to be involved exclusively with vision, in cats about one-third of the neurons in areas 17 (striate cortex), 18, and 19 are reported to be responsive to sound as well as to retinal input, and some are reported to be modulated by somatosensory input. The auditory receptive fields of these neurons are well defined in the azimuthal dimension and are roughly congruent to the visual receptive fields of the same and nearby neurons, indicating that some spatial features are conserved across modalities. The function of auditory and somatosensory input to visual cortex is not known.

At the single-neuron level, the congruence of spatiotopic maps in the optic tectum and in the visual cortex is reflected by the receptive fields (or best areas) of OR neurons, which have a strong tendency to be spatially aligned. Furthermore, at least in cat superior colliculus, maximal nonlinear enhancement appears to depend upon spatial alignment.

Structural basis of modality combination

It is useful to relate the architecture of dendritic fields of multimodal neurons to the organization of sensory input to the brain. In general, neuropil invested with one sensory modality of input is, to a greater or lesser extent, spatially segregated from neuropil invested with another; furthermore, the modality mix evident in a neuron's physiological responses reflects the modalities of the regions crossed by its dendritic arbor. This *geometric principle of modality integration* is applicable to vertebrates and invertebrates: (1) In the optic tectum, afferent terminals of different modalities are segregated by lamina (though the laminar segregation is incomplete, and there is evidence for some additional types of segregation within deeper laminae). Retinal input fibers are concentrated in the superficial layers, and auditory, infrared, electrosensory, or somatic inputs (in the appropriate species) are concentrated in deeper strata. Multisensory tectal cells are usually found to have dendritic trees that span deep and superficial layers (though there are exceptions). (2) In spinal cord, different regions of neuropil serve pain and tactile modalities, and multimodal cells ramify within both. (3) In dipteran insects, distinct regions of deutocerebral neuropil are devoted to different senses. Multimodal interneurons have dendritic patterns that sample regions corresponding to each of their modalities. The geometric principle is not rigidly applicable for all modality integrating neurons. There is in some (probably, most) cases synaptic specificity by which multisensory neurons eschew contact with some classes of processes within some parts of the neuropil that are encountered by their dendrites. Details of such *selective connectivity* remain to be discovered.

Intermodality calibration

During the course of an animal's development, neurons in multisensory regions must become appropriately connected to inputs of each modality that is to be represented (so that spatial location is properly calibrated in each sense). Studies of the superior colliculus of neonatal kittens show that functional somatosensory connections develop before birth, auditory connections just after birth, and visual ones several weeks later. This raises the questions: (1) Does each modality's map form according to its own genetic plan, independently of others? (2) Does the map formed by one set of inputs form a map that serves as a template for other senses (e.g., the somatosensory or auditory spatiotopic map may calibrate the late-developing visual map)? (3) Does the nonvisual map(s) get rearranged to match the late-developing visual map? (4) If there is rearrangement or specification of one modality's map to match another, is the rearrangement determined by a system of biochemical markers, or is it determined by experience?

The principle of *intermodality calibration* has been demonstrated in owls. The spatial receptive fields of adult owls depend on neural analysis of time and intensity differences registered at the two ears, so that plugging one ear leads to mislocalization of sounds and to misalignment of auditory and visual fields in the tectum. Owlets that are raised with an ear plug for several weeks after hatching develop relatively normal visual-auditory correspondence of their tectal spatiotopic maps. If, within a "critical period," the ear plug is removed and maps are remeasured immediately, the visual-auditory misalignment that is found reveals that the auditory map was shifted to compensate for the plug and to match the visual map. Then if the owlet experiences a period of weeks without the ear plug, the auditory map reorganizes until it once again matches the visual map.

That some experience-dependent changes can occur during mammalian development is suggested by experiments in which the eyes of kittens are occluded during rearing. They reveal an abnormally high incidence of superior collicular neurons that can be driven by auditory stimuli. These neurons also have an unusually broad laminar distribution and may even invade layers that are normally exclusively visual.

Intermodality calibration can be an ongoing process that continually modifies the interrelationship of information from diverse sense organs. The best known example is the vestibulo-ocular reflex (by which semicircular canal input counter-rotates the eyes against head motion in order to stabilize the image on the retina). Even in adult animals, this reflex is modified (calibrated) on the basis of ongoing experience by a process involving visual-vestibular integration and circuitry of the cerebellum.

Contextual modulation

A sensory message may have different meanings depending on the circumstances surrounding its origin (which are often monitored by a distinct sensory system). Senses that encode the interrelationships of the sense organs, the body, and the environment provide such contextual information.

Prominent examples are the vestibular and proprioceptive senses, which have been shown to modify visual responses in invertebrates and vertebrates. Thus, borders of the visual fields of certain crustacean interneurons shift with static orientation of the animal relative to gravity. Similarly, neurons in superior colliculus and visual cortex of mammals are systematically affected by varying degrees of static tilt of the head. Such modulation could impart a space-constant framework to visual responses (e.g., orientation-selective cells in the visual cortex, if properly modulated, might have a preferred axis that remains fixed relative to gravity). A somewhat different role for intersensory modulation is suggested by the prominent input from neck and eye proprioceptors to visual (and probably multisensory) cells of superior colliculus. The effects of these proprioceptive inputs are powerful and pervasive. In cats, passively imposed stretch of the eye muscles can excite collicular cells or can enhance or depress their visual responses. Nearly half the superficial layer neurons were reported to express one of these forms of interaction. Over 75% of neurons sampled in superficial and deep layers could be directly excited by sudden release of neck muscle tension. A similar proportion of tectospinal neurons (one source of tectal output) are excited by shocking a nerve that innervates neck musculature.

Eye muscle receptor modulation of visual responses may allow the brain to distinguish visual movement caused by motion of objects from that due to motion of the eye or might provide information to compensate neural responses for eye position or motion. Neck proprioceptive modulation might allow expression of spatial information in a body-centered coordinate frame rather than one related to the eyes, ears, or head (which can move relative to each other and to the body), though this role has not been critically evaluated.

Convergence of multiple proprioceptive inputs occurs onto cells not usually thought to have sensory function, as in those of mammalian motor cortex. Vestibular, muscle, and joint receptor input probably provide contextual information that modulates premotor commands made in response to sensory or internal signals. They may permit adjustment of effector commands to compensate for present positions or motions of limbs (or other effectors).

Nonspecific interactions

Many animals have quiescent or inattentive states. In vertebrates, a neural system associated with arousal to the attentive state (part of the reticular activating system) receives inputs from many senses. The arousal system influences neurons in almost all brain nuclei that deal with specific sensory information. Thus midbrain, thalamic, and cortical neurons primarily responsive to visual stimuli may be activated or modulated by auditory or somesthetic stimuli via nonspecific arousal mechanisms. These nonspecific intersensory interactions are characterized by sluggish changes in resting frequency of firing or in responsiveness. The changes tend not to reflect the temporal properties of the arousing stimulus, rather tending to outlast it. Phenomena like *intermodality arousal* are also seen in invertebrates. They can be expressed at quite peripheral levels. In octopus retina, responses can be modulated by tactile stimulation, presumably via efferent axons reported in optic nerve.

Mixed-mode transduction

Sensory receptors may respond with increased or decreased firing to change of more than one aspect of the environment. To the extent that they do so in the normal course of an animal's activities, this *mixed-mode transduction* will be reflected in signals sent into the central nervous system. Thus, central integration may be required if there is a need to separately identify what aspect of environmental change is responsible for a message. Examples are found among insects, crustaceans, and vertebrates, including mammals. Certain hair sensillae of insects respond both to changes of humidity and to temperature of ambient air. Neurons of crustacean statocyst organs innervating hair receptors that are responsive to substrate vibration also respond to rapid (but not slow) acceleration. The ears of some reptiles and birds can detect both airborne sound and substrate-borne vibration; it is likely that some of the ear's hair cells respond to both modes of stimulation. Some skin receptors of mammals respond both to mechanical and to thermal stimuli. In none of these cases is it known if the multiple sensory variables are separated in the brain (i.e., if demultiplexing occurs). It has been pointed out that the responses of two or more mixed-mode receptors having different mixtures of sensory variables (e.g., different ratios of sensitivity to sound and vibration) could be compared in the nervous system to reduce or eliminate the ambiguity of their signals. The problem of extracting color information from a set of photoreceptors that have overlapping spectral sensitivity curves is analogous, in many ways, to the problem of resolving mixed-mode ambiguity.

A second form of *mixed-mode transduction* involves a primary environmental variable that causes increased or decreased firing frequency, and a secondary modulating variable (usually temperature) that alters the primary response but engenders no response by itself. Thus, insect and reptile auditory receptors, crustacean statocyst neurons, fish electroreceptors, and mammalian skin vibration receptors may have thresholds, frequency vs sensitivity curves, amplitude vs response curves, and response dynamics that are altered by changes in temperature of the animal. Such *modulatory mixed-mode* transduction may play a useful role, but the adaptive value of most examples is obscure. More often *mixed-mode transduction* is likely (1) to generate a level of confusion about modality of origin that is inconsequential to the animal, or (2) to have required the

evolution of circuitry to distinguish among different possible sensory cues that could have resulted in the same signal. Thus, inhibitory modulation of spinal cord pain neurons by touch afferents may function to suppress the sensation of pain due to the activation of nociceptor by nondamaging stimuli.

Further reading

Hartline PH (1985). Multimodal integration in the brain: Combining dissimilar views of the world. In: *Modes of Communication in the Nervous System*, Brumwasser FS, Cohen M, eds. New York: John Wiley and Sons

Horn E, ed (1983): *Multimodal Convergences in Sensory Systems. Fortschr Zool* 28

Knudsen EI (1984): The role of auditory experience in the development and maintenance of sound localization. *Trend Neurosci* 7:326–330

Muscle Receptors, Mammalian

Douglas G. Stuart

Most mammalian muscles abound with sensory receptors that subserve a variety of mechanoreceptive, thermoreceptive, ergoreceptive, and nociceptive functions. However, only two receptor types, the spindle (named for its appearance) and the Golgi tendon organ (named after its discoverer), have transducing properties appropriate for a prominent role in the moment-to-moment control of muscle force and length. Their properties are emphasized here.

Distribution

Few tendon organs are in the tendons proper. Rather, they connect a small number of muscle fibers with an aponeurosis of origin or insertion, the latter being tendinous sheaths that usually extend along the length or deep into the belly of the muscle. The location of tendon organs delineates the boundaries of the muscle-fiber volume occupied by the spindles. This spindle-rich volume, whether extensive or restricted, incorporates the line of muscle pull and, in heterogenous muscles, it is also rich in fatigue-resistant muscle fibers supplied by low threshold motoneurons. As a result of these various associations, tendon organs and spindles appear optimally located for the detection of minute and sustained changes in muscle length and active force development. It is not known if this arrangement also holds in receptor-poor muscles. However, it can be inferred that in a relatively large muscle with a relatively small number of muscle receptors, the tendon organs and spindles are not evenly scattered throughout the muscle but rather are concentrated in a region where a change in mechanical status might be of particular significance.

Structure and transducer mechanisms

Tendon organs. Each tendon organ consists of collagenous fascicles, completely encased in a connective tissue capsule (length ca. 500–1500 μm; width ca. 50–150 μm). At one end of the capsule, the fascicles connect to an aponeurosis. At the other end, they connect with 5–25 muscle fibers, each usually belonging to a different motor unit. Intertwined between the fascicles are the terminations of a large, thickly myelinated sensory axon termed the Ib axon (''I'' for diameter and conduction velocity; ''b'' to distinguish it from Ia axons supplying muscle spindles). Sometimes, the capsule contains a smaller thinly myelinated (group III) axon whose function remains unknown. These structural features of tendon organs are qualitatively similar across muscles and species. However, overall dimensions (like those of spindles) may be slightly greater in larger muscles and larger species.

The tendon organ operates like a strain gauge. Forces developed by the inserting muscle fibers (whether produced actively by the nervous system or passively by muscle stretch) result in micromovements of the collagenous fascicles. In turn, these movements distort and thereby excite (depolarize) the mechanosensitive Ib axon terminations which are sufficiently sensitive to respond to the contraction of even a single muscle fiber. These depolarizations are summed at the axon's trigger zone (thought to be a single site within the capsule). If above the threshold, action potentials are initiated at the trigger zone and propagated to the central nervous system (CNS). Typically, the inserting muscle fibers belong to different motor units, with widely varying functional thresholds and force-producing properties. As a result, during active muscle contraction (be it a few milligrams or several kilograms of force) each of the excited tendon organs exhibits discharge which provides the CNS with a measure of local intramuscular force. Presumably, the summed responses of all the tendon organs provides a more accurate estimate of total muscle force than the discharge of a single tendon organ.

Muscle spindles. The spindle's structure is much more complex than that of the tendon organ. It consists of 2–12 specialized muscle fibers, encased at the mid-portion within a fusiform-shaped capsule (like that of the tendon organ). The capsule is 80–200 μm in width (it also contains a gelatinous material) and about 1–5 mm in length, as compared to the 3–20 μm width and 4–10 mm length of the encased fibers. These fibers are called intrafusal to distinguish them from surrounding longer and wider extrafusal fibers, with which they lie in parallel. The force developed by extrafusal fibers operates on the skeleton. In comparison, the force developed by intrafusal fibers is negligibly small. Nevertheless, it lengthens the intrafusal fibers' own mid-portion which, also unlike extrafusal fibers, is deficient in contractile machinery. Rather, it is replete with nuclei arranged in a bag- or chain-like fashion. These and other morphological, biochemical, and physiological differences are the basis for the intrafusal terminology bag (length 8–10 mm) and chain (4–5 mm) fibers, with the bag fibers further subdivided into 1 and 2 categories. The motor innervation of these three fiber types is by fusimotor axons, which usually innervate either bag_1 fibers or varying combinations of bag_2 and chain fibers. The motoneurons of fusimotor axons are called gamma if they supply intrafusal fibers exclusively and beta if they innervate a combination of intrafusal and extrafusal fibers as in submammalian species. Within each spinal (and brain stem) motor nucleus, beta and gamma motoneurons are randomly intermingled with alpha motoneurons, which latter cells supply extrafusal fibers exclusively. The prevalence of motoneurons in each motor nucleus is in the order: beta < gamma < alpha, with beta innervation only secure for selected mammalian muscles and species. In regard to afferent innervation, each spindle usually has a dual supply: (1) from the Ia axon; and (2) from a somewhat smaller and slower conducting myelinated axon, termed sp II to distinguish it from other group II axons supplying other types of receptor. The Ia axon (usually one per spindle) terminates within the capsule in a primary ending which consists of a group of spirals, each of which envelopes the mid-portion of each intrafusal fiber. Immediately adjacent are secondary endings (particularly on chain fibers) which consist of other spiral (and sometimes "spray") terminals, which unite to form the sp II axon (0–5 per spindle). Many spindles also receive autonomic innervation, but its function remains unknown.

The spindle's primary and secondary endings are mechanoreceptive, being stretched by elongation of the mid-portion of one or more of the fibers they innervate. The subsequent depolarization is impressed upon the axonal trigger zone which (as in the Ib axon) is within the capsule for both Ia and sp II axons. The stretch stimulus is provided by contraction of intrafusal fibers, stretch of a muscle, or a combination of both actions. The primary and secondary endings have a similar sensitivity to sustained (static) stretch but, at even the slowest of stretch velocities, the primary ending exhibits a substantially greater response. However, as stretch velocity is progressively increased, the two endings' dynamic responses increase in essentially similar degree. Indeed, neither ending provides a faithful velocity response. Rather, the primary ending seems best activated by minor irregularities in the length of its spiral endings while the secondary ending, lacking pronounced dynamic sensitivity, provides a response which seems to associate with the absolute length of its terminations.

Spindle structure is emphasized here because there are still many gaps in our understanding of its relation to function. For example, why do some but not all spindles have beta innervation and how does this innervation interact with gamma innervation? No model exists which provides a means of predicting how the spindle will behave (i.e., in terms of afferent discharge) in all experimental paradigms on the basis of its behavior in some paradigms. Similarly, during natural movements, no model can predict the nature of fusimotor discharge from measurements of muscle length and afferent discharge. At an even grosser level, physiological experiments have not yet accommodated the striking structural differences between the "typical" spindle with one bag_1, one bag_2 and four chain fibers, and others with either no bag_1 fiber, or a bag_2 fiber shared in tandem by one or more other spindles. Clearly, the structural complexity and variability of the spindle has not been matched with corresponding physiological information.

Signaling of muscle force and length

Properties of tendon organs and spindles which make them the likely muscle receptors for the signaling of muscle force and length, respectively, include: (1) a highly consistent response to moment-to-moment changes in the mechanical status of the muscle; (2) a low threshold to their respective adequate stimulus (a few milligrams for tendon organs; a few micrometers for spindles); and (3) a progressive increase in their firing rates as the strength of their respective stimulus increases to the physiological limit. In addition, the spindles have three additional properties that must play a significant role in the responses of their afferents: (1) a small-perturbation sensitivity that is relatively independent of muscle length; (2) a dual afferent-axon supply, the Ia afferents seeming best suited for the detection of small irregularities of movement and the sp II afferents possibly signaling absolute changes in muscle length; and (3) an efferent (fusimotor) control, the effects of which are difficult to generalize because they are dependent

in part on the type of spindle afferent (i.e., Ia or spII), the type of fusimotor axon (beta or gamma, static or dynamic), and the amount of change in muscle length. It is noteworthy that these statements on receptor properties hold when the afferent discharge of tendon organs and spindles is recorded in conscious humans and animals, the latter studies suggesting the further generalization that fusimotor control serves to maintain spindle afferent discharge in the mid-range of its possible extremes when the movement proceeds as planned. From this perspective, the role of the fusimotor system is to enable the spindles to report not what is going on, but how different it is from the expected.

Coexistence of spindles and tendon organs. During natural unobstructed movements, spindle and tendon organ responses from the same active muscle are sufficiently similar to raise the question as to why most muscles are provided with two low-threshold mechanoreceptors. This essentially teleological issue is seldom raised in the muscle-receptor literature. It has been argued that both receptor types are necessary to help the CNS distinguish between internal changes (e.g., fatigue) and external impediments (e.g., inertia). More speculation of this type would be valuable because it emphasizes that mechanoreceptors are designed to contribute to the control of movement.

Other muscle receptors

Muscles contain many other types of sensory receptor, but, for technical reasons, they have not been studied as systematically as spindles and tendon organs. Their relative abundance (typically supplied by > 50% of the muscle's afferentation) attest to their functional importance.

Distribution and structure. Paciniform corpuscles (with a laminated structure) may be more plentiful than generally supposed. They are always found juxtaposed to tendon organs. Other differentiated receptor types include the much rarer and randomly distributed Pacinian corpuscles (also laminated but much larger) and Ruffini-ending receptors. These three receptor types are supplied by group II and sometimes by group III (and even rarer group I) axons. The remainder of the muscle's receptors have group III and unmyelinated group IV axons, which terminate with free endings randomly throughout the muscle.

Function. Muscle receptors other than spindles and tendon organs are concerned with mechanoreceptive, thermoreceptive, ergoreceptive, and nociceptive functions. Perhaps the first three of these functions are all ergoreceptive to the extent that their input might be required for exercise-related rises in blood pressure and heart rate. But more roles also seem certain. For example, the well-known clasp knife reflex is not attributable to spindle or tendon organ input, as has been commonly supposed. Rather, other mechanoreceptors are involved. Similarly, ergoreceptive function might well involve several exercise-related modalities subserved by specific receptors with specialized central actions. What does seem clear is that about 50% of a muscle's afferentation (in the group III and, more prominently, the group IV range) subserves nociception; an understandable arrangement, since muscle damage can occur during contraction. Metabolite release, associated with extreme exercise, can excite many of these nociceptive endings; such release during less severe exercise appears not to excite muscle afferents, as was once supposed.

Further reading

Monograph

Matthews PBC) (1972): *Mammalian Muscle Receptors and Their Central Actions*. London: Arnold

Reviews

Hazan Z, Stuart DG (1984): Mammalian muscle receptors. In: *Handbook of the Spinal Cord*, vols 2 & 3, Davidoff RA, ed. New York: Marcel Dekker

Loeb GE (1984): The control and responses of mammalian muscle spindles during normally executed motor tasks. In: *Exercise and Sports Sciences Reviews*, Vol 12, Terjung RL, ed. Lexington, Massachusetts: Collamore

Comparative information

Barker D (1974): The morphology of muscle receptors. In: *Handbook of Sensory Physiology III/2, Muscle Receptors*, Hunt CC, ed. Berlin: Springer-Verlag

Proske U (1981): Properties and central actions of tendon organs. In: *Neurophysiology*, Vol 9, Porter R, ed. London: Butterworths

Proske U, Ridge RMAP (1974): Extrafusal muscle and muscle spindles in reptiles. In: *Progress in Neurobiology*, Vol 3, Pt 1, Kerkut GA, Phillis JW, eds. Oxford: Pergamon

Muscle Sense

Peter B.C. Matthews

We are all aware of where our limbs are and what they are doing, whether moving or still, and whether exerting effort or not. The origin of this muscle sense, or kinesthesis, has been debated for well over a century but, rather surprisingly, certain essentials remain unresolved. The central question is where the neural signals come from to impinge upon the sensorium (or higher sensory centers) and so lead to our conscious sensation. The ensuing question, one currently much less accessible to investigation, is how the information from different sources is compounded to produce the overall sensory experience. The obvious source for the underlying sensory input is the discharge of specific afferent receptors in muscle, tendon, joint, and skin. As long agreed, there can be no doubt that at least some of these play an essential role. It is less intuitively apparent, however, that in principle muscle sense could equally derive from the activity of the motor centers by their supplying the sensory centers with copies of the appropriate outgoing motor messages (corollary discharges). This may be illustrated by a conceptually simple, but largely disproved, example. If limb movement were to be produced by a powerful, spinally operated, follow-up length servo, the descending command would specify position, and by reading these signals the higher sensory centers could estimate the moment-to-moment position more expeditiously than if they had to await the delayed arrival of afferent feedback from the periphery. For the eyes such sensations of innervation have long been thought to exist. For the limbs the question is whether the genuine sensory message transmitted to the higher centers can be read unambiguously without the help of corollary discharges.

Classically, the joint receptors have been taken to be the principal source of sensory information leading to awareness of joint position. Early afferent recordings in animals supported this view, but later studies raised serious doubts. First, many individual receptors have been found to discharge at both extremes of movement of a hinge joint and so provide an ambiguous signal of joint position (though this might conceivably be an artifact of the experimental situation). Second, and more seriously, virtually all the receptors of a given joint may be silent when it is in the middle third or so of its physiological range of movement so that they then entirely fail to provide a signal of joint position, although the sensory centers undoubtedly continue to be informed of this. In addition, patients who have been provided with artificial joints continue to be aware of their angular position, although these lack joint receptors. The question remains whether the joint receptors do or do not contribute to position sense, when they happen to be firing, or whether they subserve quite other functions for which their apparent deficiencies are immaterial. Cutaneous afferents are also inevitably excited by the skin deformation that occurs when a nearby joint moves, but whether this contributes to muscle sense remains unknown.

The logical gap left by the dethronement of the joint receptors has now been filled by the muscle afferents, especially the muscle spindles. For a 20-year period around 1960 these had been thought to be reserved for subconscious motor control, though always previously having been believed to contribute to muscle sense. The initial evidence leading to their reinstatement was provided by studying the kinesthetic illusions that may be readily elicited by applying vibration percutaneously to human tendons to excite muscle afferents (normally 100/120 Hz by a physiotherapy vibrator). This elicits the sensory experience that the joint at which the vibrated muscle operates is moving in the direction that it would if the muscle were being stretched, even though the limb remains immobile. Afferent recordings show that vibration powerfully excites the primary endings of the muscle spindles. Thus, the occurrence of the illusion argues that when these afferents are excited normally, by muscle stretch occurring in conjunction with joint movement, their discharges must contribute to the sensation of movement; but, most interestingly, the sensation is referred to the joint rather than to the muscle. The illusion is chiefly one of continuous movement, as seems appropriate for the excitation of afferent fibers which are normally much more effectively excited by dynamic than by static stretching of a muscle. Indirect evidence suggests that the discharge of the spindle secondary endings contributes to the awareness of the static position of a limb.

Further incontrovertible evidence for a sensory role for the muscle afferents has been obtained by stretching muscles when joint and cutaneous inputs have been excluded. This can be done most directly in the course of surgery without general anesthesia, when the subject reports movement of the relevant joint although no movement actually occurs. Alternatively, selective muscle afferent activation can be achieved by moving a finger after eliminating local cutaneous and joint afferents by circulatory occlusion or by local anesthesia; the long finger muscles lie in the forearm and so are readily spared to continue to provide a muscle afferent input on moving the joint. The subject is then fully aware of movement of the otherwise insentient digit, which is a quite remarkable experience. Related experiments in subjects with anesthetized hands showed that they could continue to judge the stiffness of springs held between thumb and finger, showing that they could simultaneously assess both length and tension. This can be taken to show a sensory awareness of the signals from tendon organs, indicating tension, in addition to those from muscle spindles, indicating muscle length; certain essential controls make it unlikely that either signal could be derived solely from corollary discharges.

The acceptance of a sensory role for the spindle afferents leads almost invariably to the acceptance also of the existence of corollary discharges. This is because without their help the spindle message would seem to be unreadable by the higher centers, since an increase in Ia firing can be due either to muscle stretch or to an increase in fusimotor discharge. As these two situations do not seem to be confounded, one is driven to conclude that the motor centers tell the sensory centers when, and to what extent, the fusimotor system is being activated. The complex details of such processing still entirely elude us except for a striking asymmetry of the roles of the sensory input and the corollary message. The afferent input can elicit a sensory experience on its own, in the absence of a corollary signal, since moving or vibrating a passive muscle leads to sensation. But a corollary discharge occurring in the absence of a changing sensory input produces no subjective experience of movement. This absence of effect is found when the subject attempts to move a limb that has been paralyzed by local anesthetic (as in spinal anesthesia) or by local curarization. The most convincing situation is when the locally anesthetized subject fails to perceive a movement that he actually makes, which can be confidently presumed to be accompanied by the usual corollary discharge. This arises, it is believed, when the afferents have been paralyzed in advance of the efferents. Thus, for the sense of movement and position, corollary discharges seem cast in the role of providing an interpretative signal to allow proper decoding of the sensory message, rather than having a direct sensory action on their own.

However, a powerful case can be developed that other corollary discharges do lead directly to another sensory experience, namely, that of muscular effort. Moreover, these corollaries seem to be routinely employed by our higher sensory centers when we are judging the heaviness of a lifted object. The signal from tendon organs on force, though available, does not appear to be used for this in at least some perfectly normal situations. The perceived heaviness of an object is increased when the strength of the muscle acting upon it is weakened by fatigue or partial local curarization, thereby showing that it is the motor command rather than the actual force developed that is responsible for the sensory experience. Some higher level command must be being monitored, and not just the discharge of the relevant motoneurons, since when these are influenced reflexly, as by the tonic vibration reflex, the effort required to support a given weight is decreased or increased depending upon whether the reflex effect is excitatory or inhibitory. Of wider interest is the observation that after a stroke, including one apparently affecting just the motor system, the patient immediately complains of the great effort involved in exerting even a small force and of the great weight of everyday objects. All these effects can be demonstrated objectively by matching the perceived heaviness of objects held bilaterally. However, agreement on interpretation has yet to be reached among all those concerned. Finally, it should be noted that to postulate corollary discharges is to say nothing new anatomically, since a wealth of such recurrent pathways are known; the physiological problem is to understand the type of information that they convey and how it is used functionally.

Further reading

Matthews PBC (1977): Muscle afferents and kinaesthesia. *Br Med Bull* 33:137–142

Matthews PBC (1982): Where does Sherrington's "muscular sense" originate? Muscles, joints, corollary discharges? *Ann Rev Neurosci* 5:189–218

McCloskey DI (1978): Kinesthetic sensibility. *Physiol Rev* 58:763–820

McCloskey DI (1981): Corollary discharges and motor commands. In: *Handbook of Physiology, The Nervous System III, Motor Control*, Brooks VB, ed. Bethesda: American Physiological Society

Noci-Reception, Nociceptors, and Pain

Edward Perl

Animal tissues are constantly under the influence of their environment. That environment, of course, is both external and internal to the body. By and large afferent nervous structures have adapted to indicate the state of most tissues and some of the usual changes occurring in them. Because mammalian tissues are fragile compared to objects or situations in the natural world, they can be injured by anything from sticks and stones to fires and infections. These injurious events or interactions with tissues have been grouped under the term noxious. Tissue injury usually causes human beings pain, and, in fact, the term noxious was first applied to stimuli in an attempt to define a common basis for the variety of situations evoking pain. However, it must be kept in mind that pain is a perceptual reaction, a sensation. Pain is thus subjective and only one of a number of reactions capable of being initiated by noxious stimuli. Detection of noxious events, or noci-reception, has obvious protective value, and in the normal individual, pain is part of a complex of protective reactions associated with noci-reception.

Sense organs for tissue injury

The primary afferent fibers of peripheral nerves (extensions of neurons of dorsal root or cranial nerve ganglia) are heterogeneous, functionally and structurally. Their terminations in target tissues, by themselves or combined with other specialized cells, transduce events and circumstances into nerve impulses. Primary afferent neurons (sensory units, receptors) may be classified on several grounds, including the termination tissue, the effective stimuli, the diameters or their afferent conduction velocities of their fibers, and the structural or chemical features of their cell bodies and their peripheral and central terminals. On these bases each category of primary afferent neuron presents a unique constellation of properties.

When exposed to qualitatively and quantitatively different stimuli, a primary afferent neuron responds with a considerable but by no means absolute selectivity. A large proportion of primary afferent neurons are most effectively excited by innocuous stimuli, as, for example, moving contact with hairs or skin, cutaneous temperature changes, stretch of muscles, movement of joints, and normal intestinal peristalsis. In terms of integrity of the tissue, they are low in threshold for at least certain stimuli.

In most peripheral tissues a subset of primary afferent neurons, in contrast to low-threshold units, have substantially elevated thresholds for all ordinary stimuli. Many of the high-threshold sensory units are only excited by stimuli that cause overt damage to the tissue. However, regardless of stimulus intensity required to initiate activity, the high-threshold class is characterized by an increased response to progressively more intense noxious stimuli. Therefore, by both threshold and the response to different levels of noxious stimulation, high-threshold sense organs unambiguously signal the presence of noxious situations; they are noxious stimulus receptors, i.e., noci-receptors or nociceptors.

Low-threshold sense organs respond quite differently when exposed to tissue-damaging stimulation. First, their maximal discharge is elicited by a type and an intensity of stimulus that is not tissue damaging. Second, many are actually inactivated by the kind of noxious events exciting nociceptors and give neither a greater discharge nor an unusual pattern of response to tissue-damaging events. Documentation of these differences in behavior of low- and high-threshold somatic sense organs has laid to rest theories about the origin of pain that presumed pain-causing events to initiate either an excessive discharge or a special pattern of activity in low-threshold elements.

Varieties of nociceptors

Nociceptors have afferent fibers conducting from the range

Table 1. Varieties of Cutaneous Nociceptors

| Type | Afferent Fiber | Responsiveness to Noxious Stimuli | | | Receptive Field |
		Mechanical	Thermal	Chemical	
High-threshold mechanoreceptor	myelinated 40–5 m/sec (Aβ-Aδ)	+ + to + + +	none or delayed (heat)	− to ±	multiple distinct point-like areas (<0.1 mm^2)
Polymodal nociceptor	unmyelinated (C) 1–0.3 m/sec	+ + +	+ + + +	+ + to + + +	usually one small area (1–5 mm^2)
Heat mechanoreceptor	myelinated (Aδ) 15–4 m/sec	+	+ + +	?	small area
Cold mechanoreceptor	unmyelinated (C) < 2 m/sec	+	+ + +	?	small area

associated with the medium to thinnest myelinated fibers (~40–4 m/sec) down to velocities associated with the thinnest unmyelinated fibers (1.0–0.3 m/sec). Their cell bodies in the dorsal root ganglia vary from those of intermediate diameter (30–40 μm) to some of the smallest (15 μm). While actual values differ by species, the range relative to the overall population of primary afferent neurons appears to hold for most mammals, including human beings. It is important to note that both mechanoreceptive and thermoreceptive units selectively responsive to innocuous stimulation also are present in the same range of fiber and ganglionic cell body sizes.

There are distinctive functional differences within the nociceptor population, some of which correlate with morphologically or chemically identifiable features other than cell body or fiber size. More than one type of nociceptor has been established for skin, muscle, and joints, and probably also exists in other tissues. Since most is known about the features of cutaneous nociceptors, they are used as examples. In primate skin, two distinct types of cutaneous nociceptors are readily demonstrable, with evidence for two additional varieties. General features of cutaneous nociceptors are depicted in Table 1.

The first two entries in Table 1, myelinated fiber high-threshold mechanoreceptors (mechanical nociceptors) and unmyelinated (C) fiber polymodal nociceptors, describe nociceptors with similar characteristics in the cutaneous innervation of many mammals including *Homo sapiens*. Evidence favoring the existence of the other two types is less extensive but indicates that in some species or skin regions they represent unique classes of nociceptive sense organs. Skin and joint also have been shown to have nociceptors with myelinated as well as unmyelinated afferent fibers, although functional and other differences in the subcutaneous nociceptors need further documentation.

Many dorsal root ganglion neurons with unmyelinated fibers give evidence of the presence of substantial quantities of one or more peptides that have been implicated in cell-to-cell communication either as hormones or as putative synaptic mediators (e.g., substance P, somatostatin, vasoactive intestinal peptide, calcitonin gene-related peptide, and others). Purely on the basis of the number of such neurons present in the ganglia, a connection between nociceptors and some of these peptides could be hypothesized. Another, and somewhat less circumstantial connection to nociception was established by the use of the selective neurotoxin capsaicin, which when given parenterally in young rodents is associated with decreased reactivity to some noxious stimuli. Such treatment causes peptide decreases in dorsal root ganglia neurons and in the terminations of fine primary fibers in the spinal cord.

Synaptic transmitters for nociceptors

The synaptic articulation of nociceptors to central neurons is only partially understood. There is direct evidence that nociceptors with myelinated and with unmyelinated fibers terminate in the superficial part of the spinal dorsal horn (marginal zone and substantia gelatinosa) and the region near the spinal central canal. Myelinated fibers from the cutaneous and subcutaneous mechanical nociceptors have also been traced to the neck of the dorsal horn (lamina V).

Unmyelinated primary afferent fibers giving immunocytochemical reactions for peptides in dense-core vesicles also regularly contain small round (clear) vesicles whose contents lack the same reaction. Synaptic enlargements of myelinated fibers from identified nociceptors contain small round vesicles common to excitatory presynaptic terminals but few, if any,

of the dense-core vesicles thought to be associated with certain peptides. The putative peptide mediator substance P has been reported to selectively excite spinal neurons in the vicinity of the central terminals of nociceptors, but these effects developed more slowly and lasted longer than other candidate synaptic transmitters and synaptic potentials in neurons of the region. In sum, it appears likely that rapid synaptic transfer from nociceptors to spinal neurons is mediated by undetermined chemical agent(s), possibly amino acids, and that for some nociceptors with unmyelinated fibers peptides may play a complementary or modulatory role at the same synapses.

Transduction of peripheral stimulation by nociceptors

At present there is only circumstantial evidence on the structure and mechanisms of transduction of the peripheral terminals of nociceptive primary neurons. The proposal made by von Frey in the 1890s that the sense organs for pain are "free nerve endings" is still commonly mentioned. This idea derived from comparisons of the relative number of nerve endings viewed in light microscopic histological examinations of the skin and the density of pain spots observed in psychophysical studies of skin sensibility. Recent studies (using ultrastructural analyses on skin areas underlying the circumscribed receptive spots of cutaneous mechanical nociceptors with myelinated fibers) have demonstrated unique and unusual nerve terminations, enveloped by an extension of the Schwann cell surrounding the parent fiber in the dermis, between keratinized cells of the epidermis. This observation suggests that, in fact, nociceptor endings may not be "bare" and that their elevated thresholds could result from an insulation from the local environment by nonneural structures.

Chemical intermediaries between pain-causing stimulation of peripheral tissue and the activation of afferent nerve fibers have been advocated for years. Certain substances, found in or produced by tissues when injured, evoke pain-like reactions when introduced parenterally or intraarterially. These include histamine, hydrogen and potassium ions, bradykinin and other kinins, prostaglandins, and serotonin; each has been suggested as a peripheral mediator for pain-causing neural activity. Tissue injury and infections are associated with a tenderness (hyperalgesia) that usually extends some distance beyond the actual region of damage, a feature readily recognized in skin. Hyperalgesia in the region immediately adjacent to a place of injury has been attributed to chemical agents diffusing from the point of damage. Pain can be suppressed (analgesia) by pharmacologic agents acting peripherally on particular substances. Thus diffusible chemical agents may be involved in the peripheral activation or responsiveness of certain nociceptors.

Like most sense organs, nociceptors exhibit fatigue when repeatedly activated. Some of them, however, express an almost opposite characteristic. Cutaneous C-fiber polymodal nociceptors often develop ongoing (background) discharge and enhanced responsiveness after mild to moderate tissue damage in the immediate vicinity of their receptive fields. These changes have been labeled sensitization. These two features of sensitization, the evolution of ongoing discharge and increased responsiveness, may be due to somewhat different mechanisms. Sensitization is manifested by a lowering of threshold, an increased discharge for a given stimulus, or both. It begins in a few minutes and lasts for hours after noxious stimuli. Some cutaneous mechanical nociceptors show a similar (though usually less marked) sensitization, and, in fact, may become responsive to noxious heat as a consequence. Sensitization with some equivalent features also has been noted for muscle and testicular nociceptors. The factors producing sensitization spread from the point of damage or stimulation, possibly as the result of diffusion of a substance. The process of inflammation initiated by tissue damage is presumed to involve chemical intermediaries. It is likely that a relationship exists between nociceptor sensitization and some chemical mediators associated with the inflammatory process.

Some nociceptors appear to be involved in the vasodilation produced by antidromic activation of afferent fibers. It has been proposed that this vasodilation is produced through the release of chemical substances by unmyelinated sensory fibers. The hypothesis relating the local vasodilation to nociceptors suggests that a noxious stimulus activating some terminals of an afferent fiber initiates action potentials conducted not only centrally, but also distally at branch points to excite branches and terminals outside the stimulated area. Therefore, the converse of what occurs for sensitization may apply to inflammation; the neural apparatus sensing noxious stimuli may contribute to local changes associated with inflammation and tissue repair.

Relationship of nociceptor activity to pain

For pain to be felt in the normal manner, conduction must be intact in both the thinly myelinated and the unmyelinated fibers innervating the stimulated tissue. This was established in classical experiments during which different subsets of the spectrum of fibers in peripheral nerve were selectively excited or reversibly blocked using combinations of electrical stimulation, pressure, and local anesthetics applied to the nerve. This conclusion is supported by the observations that persons lacking most of the thin afferent fibers in peripheral nerve do not feel pain on damaging stimulation and have evidence of unnoted injury. Since afferent fibers from nociceptors represent a substantial proportion of the thin fibers of peripheral nerve, these findings suggest that nociceptor signals are crucial for pain to result from peripheral stimulation.

The attributes of stimuli evoking pain and those initiating activity in particular types of nociceptors have been shown to be closely similar in psychophysical experiments. However, the most conclusive evidence on the relationship between nociceptor activity and pain has come from artificial stimulation of functionally identified single primary afferent fibers. In these experiments, transcutaneous microelectrodes placed to record activity from single afferent fibers were used to determine their functional properties. The recording electrode was then employed to electrically stimulate the same fibers without disturbing peripheral tissue. Excitation of the afferent fibers from

nociceptors was found to elicit either pricking or unequivocal pain and in some instances itch but no other sensation. On the other hand, excitation of the afferent fibers from the various low-threshold sense organs of the same nerve produced other sensory experiences, but not pain.

Projection of activity from nociceptors in the central nervous system

Selectivity for noxious stimulation on the part of sense organs does not by itself describe the remainder of the neural apparatus for pain; however, the relationships between nociceptor afferent activity and pain have important implications for aspects of the central pathways related to nociception and pain. The specificity of sensory experience obtained by stimulation of the afferent fibers of nociceptors in isolation is difficult to explain by their projection to other than more or less dedicated pathways ascending to the centers responsible for perception. Nevertheless, the presence of dedicated pathways does not preclude other arrangements in relation to noci-reception. Therefore, the fact that nociceptors have been established to connect to both relatively dedicated and broadly responsive neurons should not be surprising. A pathway consisting of a chain of neurons, selectively activated by noxious stimuli, exists from the spinal cord to the thalamus and from there to the somatosensory region of the cerebral cortex. One part of this chain seems to be activated principally by the myelinated nociceptors, and other linkages involve input from the unmyelinated fiber nociceptors. These dedicated pathways are influenced by other incoming and intrinsic activity. The role of parallel pathways receiving excitation from both nociceptors and other peripheral sense organs to the same and other higher structures in normal or abnormal pain mechanisms is less clear, although their possible role in these functions should not be ignored.

Further reading

Burgess PR, Perl ER (1973): Cutaneous mechanoreceptors and nociceptors. In: *Handbook of Sensory Physiology, Vol II: Somatosensory System*, Iggo A, ed. Berlin, Heidelberg, New York: Springer-Verlag

Dallenbach KM (1939): Pain: History and present status. *Am J Psychol* 52:331–347

Keele KD (1957): *Anatomies of Pain*. Springfield, Ill: Charles C Thomas; Oxford: Blackwell Scientific Publications

Perl ER (1984): Pain and nociception. In: *Handbook of Physiology: The Nervous System III*. Brookhart JM, Mountcastle VB, Darian-Smith I, eds. Bethesda: American Physiological Society

Olfaction

Maxwell M. Mozell

Olfaction is the sensory process that in response to chemical stimuli gives rise to those sensations called odors. The primary olfactory receptive area is that region of the nose subserved by the olfactory nerve. Although the chemoreceptive endings and neural projections of the olfactory nerve are basic to the sensing of odors, other cranial nerves are also involved, namely, the trigeminal, glossopharyngeal, and vagus. These additional cranial nerves possess at least some chemoreceptive endings which line the respiratory tract at different levels (nose, pharynx, larynx) and together define the accessory areas of olfaction. These accessory areas, especially that of the trigeminal nerve, give rise to the pungent or irritating quality often experienced as part of an odor sensation. It is not yet certain, however, that these nociceptive-like qualities are the only contribution that the accessory areas give to the total odor sensation. Nonetheless, the existence of accessory olfactory areas in addition to the primary area requires caution in testing and evaluating olfactory function since, for instance, an odorant might be detected through the trigeminal input even though the olfactory receptors are unresponsive.

Another view is to restrict the definition of olfaction to the input of the olfactory nerve alone and refer to the accessory inputs of the trigeminal, glossopharyngeal, and vagus nerves as the common chemical sense. These contrasting views highlight how difficult it is to characterize olfaction differentially from other chemical senses. For instance, many of the airborne chemicals which excite the olfactory nerve also excite, perhaps at higher intensity levels, the other chemosensitive cranial nerves. Furthermore, in aquatic animals olfactory stimuli are not even airborne, but instead are dispersed in aqueous solution making olfaction, in this regard, indistinguishable from gustation. Neither, in view of these aquatic animals, can the distinction be made that olfaction responds to molecules coming from distant sources whereas gustation seems more like a contact sense. With the stimuli for both senses dispersed in the animal's aqueous environment, each can be considered both a contact and a distance sense. Therefore, to distinguish olfaction from the other chemical senses, one often falls back upon such morphological hallmarks as the structure of the receptor cells, their positions in the nose, and the pathways of their central projections. However, these hallmarks do not characterize olfaction in subvertebrate phyla such as in insects where the morphology is quite different. For instance, the receptor cells in insects are supported on antennae.

In humans the olfactory mucosa (i.e., the patch of nasal mucosa bearing the olfactory receptor cells) lies about 7 cm up and into the nose from the external nares (Fig. 1). It covers an area of approximately 1 cm^2 (exaggerated in Fig. 1) lining the lateral, medial, and superior walls of the uppermost passageway in the nasal cavity. This passageway, often called the olfactory cleft or olfactory slit, is only 1 mm across, thus presenting to the incoming odorant molecules a rather constricted airway which is much taller than it is wide. Perhaps this configuration, by passing the molecules through a narrow constriction between tall mucosa-lined walls, increases the possibility that any given molecule will contact the olfactory mucosal surface.

The olfactory mucosa has several layers which, from surface to base, include a mucus covering, a pseudostratified columnar epithelium, and a lamina propria (Fig. 2) which rests on underlying bone or cartilage. The epithelium contains at least three cell types (receptor cells, supporting cells, basal cells) and perhaps, as recently reported in humans, a fourth type (microvillar cells). The olfactory cells are bipolar serving both as the transducing end organ and as the primary fiber projecting centrally to the olfactory bulb. A rather unique property of

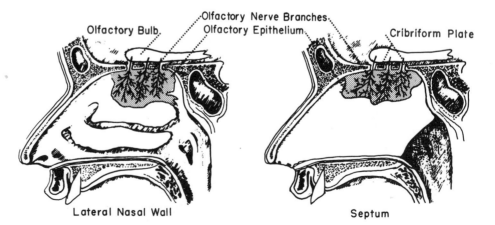

Figure 1. A sagittal section through the human nose showing the olfaction-related morphology on the lateral nasal wall and on the septum. The actual area of the olfactory epithelium is exaggerated somewhat to better depict its neural innervation.

Olfactory Bulb.

Olfactory Nerve Branches.
Olfactory Epithelium.

Cribriform Plate

Lateral Nasal Wall

Septum

the olfactory receptor cell in contrast to other neurons is that in a cycle of about 35 days it can completely degenerate and then be replaced all the way from the mucosa to the bulb by a new cell differentiating from the basal cell population. Perhaps this turnover, which appears to continue throughout the animal's lifetime, relates to the interaction of the receptor cells with odorant molecules having various toxicity levels. At any rate, any conceptualization of the olfactory process at the mucosal level must take into consideration this dynamic changeover of receptor cells.

The distal or dendritic end of the receptor cells (the olfactory rod) is about 1 μ in diameter and, in humans, about 40–50 μ long. It protrudes (Fig. 2) into the overlying mucus as a swelling called the olfactory knob which bears a number of cilia (10–30 in humans) trailing out 100–150 μ in the mucus. Studies evaluating the effects of the removal and regrowth of the cilia upon the response of the receptor cells to odorants suggest that the proximal end of these cilia are a major loci for the odorant-receptor cell interaction leading to transduction. On the other hand, cilia do not appear essential for odorant transduction. There is some suggestion that the olfactory knob itself might respond to odorants, and in some species the receptor cells do not bear cilia. Furthermore, the receptor cells lining the vomeronasal organ (a blind sac structure found opening into the nasal passageways in members of many species including some humans) bear microvilli rather than cilia but still appear to respond to certain odorants. In particular they respond, as in some rodents, to pheromones, i.e., chemicals released and received as signals between conspecifics.

There might also be nonciliated receptor cells in the human olfactory mucosa which are about one-tenth as prevalent as the ciliated receptor cells. Instead of cilia this cell, called the microvillar cell, bears 75–100 microvilli. It is not yet certain that this is truly a receptor cell; the major support for this assertion is, at present, the presence of what appears to be an axon-like process.

Although human receptor cell cilia do not seem to incorpo-

rate the components necessary for motility, the cilia in subhuman species do appear motile during the early stages of growth, but as length increases, this motility decreases. In studies following receptor cell regrowth after experimentally induced degeneration, the cilia seem to lose their motility at about the same time the axons reach the bulb. Moreover, whereas prior to these anatomical events the receptor cells seem to respond to most of the test odorants, they now become more selective.

The olfactory rods of the receptor cells are surrounded by supporting (sustentacular) cells (Fig. 2), an arrangement which separates the receptor cells from each other although occasional direct contacts between them do exist. With the possible exception of humans, the data from most of the species studied indicate that the supporting cells, which bear microvilli at their distal ends, are major contributors to the mucus layer blanketing the epithelium. Another major contribution to the mucus layer is secreted in both humans and subhuman species through the ducts of the many glands of Bowman located in the lamina propria.

The mucus could be a major factor in the olfactory process since the odorant molecules must diffuse through it to reach, at least to some degree, the transduction loci. In addition, the molecules of different odorants will interact with the mucus differently depending upon their physicochemical properties. For instance, the concentration of the molecules of any given odorant at the receptor cells depends upon their partitioning between the mucus and the air phase transporting them. The magnitude of receptor cell excitation will then further depend, in part, upon the attraction of the odorant molecules to the mucus relative to their attraction to the receptor cells. The air/mucus partitioning can also affect the spatial distribution of the odorant molecules along the mucosal surface. That is, the more the partitioning favors the mucus, the more the molecules, as they pass over the mucosa, sorb to the early loci in their flow path and the smaller their number reaching the more distant loci. This has implications for quality coding since it

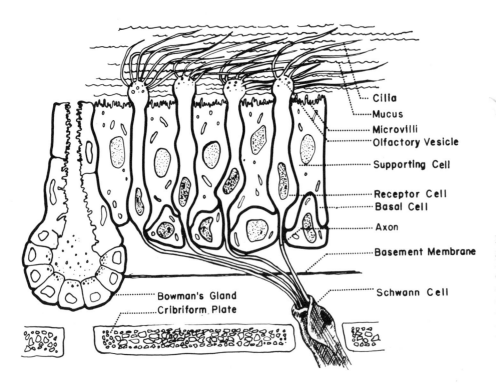

Cilia
Mucus
Microvilli
Olfactory Vesicle
Supporting Cell
Receptor Cell
Basal Cell
Axon
Basement Membrane
Schwann Cell
Bowman's Gland
Cribriform Plate

Figure 2. A schematic diagram of the generalized vertebrate olfactory mucosa. In humans a fourth type of cell, the microvillar cell, could be added. There is actually a close apposition between the Schwann cells and the supporting cells as they mutually surround the dendritic and axonal processes of the receptor cells. The exaggerated separation shown here is to better depict the course of the unmyelinated axons.

seems to play a role in producing different odorant-dependent molecular distributions across the mucosa which, in turn, appear to give to each odorant a characteristic pattern of mucosal activity. There are also implications for quantity coding. An odorant might access the receptor cells at a higher concentration than would be expected from its concentration in the air due to the piling up of its molecules onto a relatively small region of the mucosa.

The mucus may also be important to olfaction by affording a pathway (along with desorption back into the airstream and uptake by the circulatory system) for the removal of odorant molecules from the mucosa. The mucus constantly washes the mucosal surface and in humans nasal mucus has been observed to travel 7 mm/min. It is known that the molecules of different odorants are removed at different rates, but what effect this has upon olfactory perception, especially upon adaptation, is not yet understood.

The already impressive number of receptor cells in humans ($30,000/mm^2$ with a total of 6 million on one side of the nose) is dwarfed by the estimated number in the German shepherd ($120,000/mm^2$ with a one side total of 2 billion). One historically important but experimentally unsubstantiated interpretation of such huge numbers is that they reflect the matching of specifically tuned receptor cells for each of the vast number of discriminably different odorants. Another interpretation still given credibility is that they give necessary redundancy in an otherwise noisy system. Yet another interpretation is that they amplify certain aspects of the stimulus by summation of their inputs upon a much smaller number of postsynaptic neurons. Finally, this immense number of densely packed receptor cells might give a large, finely grained, surface area having a number of different regional sensitivities upon which the incoming molecules of different odorants can sorb in different spatial patterns. In this regard, the German shepherd's unilateral olfactory cilia have an estimated combined surface area of 7.85 m^2 (84.47 ft^2).

The thin (0.2 μ in diameter) axons of the receptor cells come together to form larger and larger bundles which course through the lamina propria and enter the cranium to reach the olfactory bulb via a series of small perforations (foramina) in the cribriform plate (Figs. 1 and 2). Although these axons are individually unmyelinated, they are ensheathed in tightly packed groups by Schwann cell plasma membranes. This arrangement raises the possibility of "crosstalk" between receptor cells as their axons project centrally.

The often repeated assertion that olfaction is very sensitive seems supported by such observed uses of olfactory cues as the male moth finding his mate 2.5 miles away and the adult salmon swimming inland from the sea to the place where he was spawned. Although rather impressive, a more quantitative context puts olfactory sensitivity in clearer perspective. For instance, in one study the human olfactory threshold for ethyl mercaptan was 1.6×10^{-2} mol/liter, a seemingly low concentration. However, an average human sniff (200–250 ml) of this concentration still contains for one side of the nose 1.08×10^{11} molecules which, relative to the number of receptor cells (6×10^6), is not small. Of course, these relative numbers vary with the odorant and the species. In the German shepherd the receptor cells outnumbered the molecules for one odorant at threshold.

In evaluating these figures, note that they represent the number of ethyl mercaptan molecules entering one naris, but the number actually contacting the walls of the olfactory cleft must be less for the following reasons: (1) only 5–20% of the air carrying the odorant molecules into the nose is directed toward the olfactory cleft, and (2) of these molecules many

are sorbed by nonolfactory mucosa before reaching the cleft or (3) pass through the cleft without striking its walls. Thus, for mercaptans only about 2% of the incoming molecules actually contact the walls, and it is a further fraction of these that finally reach and stimulate the receptor cells. These numbers differ, of course, for odorants having different physico-chemical properties.

Models of the human nose have indicated that by increasing flow rate more of the incoming air is directed toward the olfactory cleft. Thus, sniff vigor might be expected to influence olfactory perception. The experiments testing this expectation in humans are generally negative, but there is disagreement. However, in subhuman species the character of the sniff might be manipulated to control the passage of odorants across the mucosa. This would give the olfactory system a self-initiated ability to explore the stimulus analogous to the fingers manipulating objects held in the hand. Moreover, as shown in several subhuman vertebrates, the discharge magnitude of the olfactory nerve depends upon a number of sniff-defining variables, e.g., volume, duration, flow rate, number of odorant molecules, and concentration.

For much of the animal kingdom olfaction is basic to the maintenance of life. It is essential for finding prey, and it is a first line of defense against predators. It is, as noted earlier, used in communication. The male dog, for instance, marks his territory with urine, and when the skin of certain types of fish is damaged, it releases substances to alarm nearby conspecifics. Many studies with many species have shown a dependency of sexual behavior upon olfactory cues. Male hamsters display mating activity, even with other males, when presented with vaginal discharge from a receptive female. In many animals sexual dysfunction and even retarded development of the sex organs results when the olfactory process is compromised. Although not as extensively documented, a relation between olfaction and sex seems likely in humans. For instance, at least for some odorants, olfactory acuity in women seems better at ovulation than during menstruation, and there is some indication that olfactory cues among women can synchronize their menstrual cycles.

For civilized humans olfaction plays less of a life-and-death role than for lower animals, although there are times when the detection of such stimuli as smoke, gas, or decaying food prevents bodily harm. Instead, civilized society seems to emphasize the impressive hedonic effect of olfaction. As examples of the positive affect, people add seasonings to their foods, perfumes to their bodies, and incense to their homes. The importance of the negative effect is shown by the plethora of commercial products for use against objectionable odors. Although such firms could not exist if there were not at least some consensus for the hedonic effect of at least some odorants, people do vary considerably in their odor preferences. These preferences depend upon such variables as age, sex, socioethnic background, and previous odor experiences.

One instance in which olfaction plays a major role is, perhaps surprisingly, in the recognition of tastes. Much of what people think they taste they actually smell, and about 80% of the people coming to chemosensory clinics complaining of taste problems are actually diagnosed as having olfactory dysfunctions. In one study asking subjects to identify 21 common food substances placed on the tongue, there was a decrease from an average of 60% correct to an average of 10% correct when the nasal cavity was made inaccessible to any vapor phase molecules given off by the substances. Even coffee and chocolate, which were correctly identified by over 90% of the subjects when the nose was accessible, were not identified correctly by any subject when the nose was made inaccessible.

This is not to say that olfaction is the sole contributor to the appreciation of flavors since the perceived magnitude of a flavor, as distinguished from its identification, has been shown to depend upon some combination of both the olfactory and gustatory inputs.

Further reading

Doty RL (1979): A review of olfactory dysfunctions in man. *Am J Otolaryngol* 1:57–79

Engen T (1982): *The Perception of Odors*. New York: Academic Press

Mozell MM, Hornung DE, Sheehe PR, Kurtz DB (1984): What should be controlled in studies of smell? In: *Clinical Measurement of Taste and Smell,* Meiselman H, Rivlin R, eds. pp. 154–169 New York: Macmillan Publishing Company

Ottoson D (1983): Olfaction, In: *Physiology of the Nervous System,* New York: Oxford University Press

Olfaction, Insect

Dietrich Schneider

Insect olfaction, like vertebrate olfaction, is a morphologically and physiologically well-defined sensory modality: porous, hollow, cuticular structures (hairs, pegs, or plates) are innervated by the dendrites of sensory nerve cells specialized for the perception of airborne chemicals. Such receptor miniorgans (sensilla) are composed of (1) the formative cells that secrete the cuticle surface of the organ and (2) one to many receptor cells with their distal dendritic processes ($\emptyset$ 0.1–0.5 µm) and neurites ($\emptyset$ 0.1–0.5 µm) that connect directly to central neurons. The dendrites are unbranched or branched, contain in their finest form a minimum of one microtubule, and are bathed by the receptor (sensillum) lymph. The receptor cells respond with medium to high sensitivity and specificity to smaller or wider ranges of chemical stimuli. In terrestrial insects, the perception of gaseous (olfactory) stimuli is clearly distinguished from gustation of mostly aqueous (salt, sugar, etc.) stimuli of much higher concentration reaching the dendrite through the tip opening of specialized taste hair sensilla. Olfactory receptors of "amphibious" insects (e.g., water beetles) respond to the same stimuli (fatty acids) in air and under water, like the nasal receptors of the salamander.

Insect olfactory sensilla are concentrated on the antennae and vary in shape and number (few to 10^5), with up to 0.5×10^6 receptor cells. Controversy over localization of insect odor receptors ended with successful electrophysiological recordings of odor responses from the antennae in the 1950s. But humans have known of the keen olfactory power of insects since at least as early as Aristotle.

The cuticle of an olfactory hair sensillum (Fig. 1) is penetrated by microscopic pores ($\emptyset$ 10 nm; 1–100/µm² of the hair surface) that are continuous with pore tubules running through the receptor lymph space and often ending on the dendritic cell membrane. In odor sensilla without pore tubules, the cuticle is interrupted by gaps filled with electron-dense material. The system of pore openings is only a small fraction of the surface of the sensillum, but access of odor molecules to the receptive dendrite is assured and dehydration of these delicate sensilla (hair $\emptyset$ 1–5 µm) is prevented. In the transduction process, odor molecules are absorbed on the sensillum cuticle and reach the pore or gap and eventually the dendrite by diffusion. Here, the essential and selective weak binding process between the signal and the receptor molecule comes into play, and eventually ion channels elicit electric charge shifts. If the pore tubules do not reach the dendrite (or the sensillum has no tubules), there must be a direct or carrier-assisted transfer of odor molecules through the lymph space to the dendrite. Recordings from single receptor cells with extracellular electrodes revealed elementary receptor potentials (adding up to generator potentials) and traveling nerve impulses. Minimum reaction times for these potentials are 10–20 msec, but depending upon the stimulus strength, they may be much longer, indicating the complicated stimulus transport from the sensillum surface to the dendrite. Inactivation of the absorbed stimulus molecules, indicated by the decline of the generator potential after the end of the stimulus, is a rapid enzymatic process. Another slower enzymatic process that degrades odor molecules on and in the antenna and on body scales might be important to counteract contamination.

A simple approach to some stimulus response characteristics is the electroantennogram (EAG), an overall, odor-induced generator potential of the whole antenna. The EAG is particularly useful if a given antenna has many identically responding receptor cells for a species-specific odor, or pheromone. EAG recordings combined with gas-chromatography of pheromone gland extracts, fascilitate pheromone identification.

The antennal pheromone-perceiving system of the male silkmoth (Bombyx) is optimized for molecule catching (on the long hair sensilla), molecule transfer, and extreme sensitivity. Its receptor cells respond to single molecule impacts with one impulse each, but hundreds of such impacts per second are needed to overcome the receptor cell noise and to alert the moth. The quantitative range of the receptor cells covers several decadic steps. After strong stimuli, the cells adapt and may need many minutes to recover. The qualitative range (spectrum) of effective odor stimuli for pheromone receptors is narrow; geometric and even optical isomers of the pheromone are 100–1000 times less effective. This high specificity indicates that such cells possess a uniform receptor population on the dendritic surface. Receptor cells with equal spectra of effective odor stimuli, which are regularly found in identically structured sensilla in great numbers on antennae, were named "odor specialists." Other less-sensitive insect odor receptor cells such as pegs or honeybee pore plates respond (although still individually specific) to a variety of odorants, indicating that they possess more than one type of receptor in their dendrites. Relatively few cells of this response type on one antenna share the same reaction spectrum, but the spectra of neighboring cells overlap. This cell population might enable the insect to discriminate many odorants (best known from honeybee behavior studies) and the cells were, therefore, called "odor generalists." The specialist-generalist terms have practical value, but all odor receptor cells have their individual specificity. Interestingly, the great number of pheromone odor (specialist) receptor cells for the female attractant is restricted to males in moths and cockroaches; females lack receptors for their male-attracting odor.

Little is known of the signal processing from the antenal receptor cells (with their cell somata near the base of the sensillum) to the olfactory brain, the deutocerebrum, which the axons all reach without fusions. As in other sensory systems, the number of receptors neurons is much higher (> 100 times) than the number of interneurons. Prominent deutocerebral structures are glomeruli (Fig. 2), analogous to the vertebrate olfactory lobe. In male but not female moths

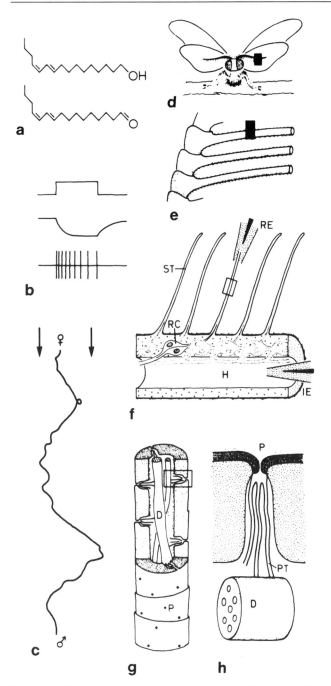

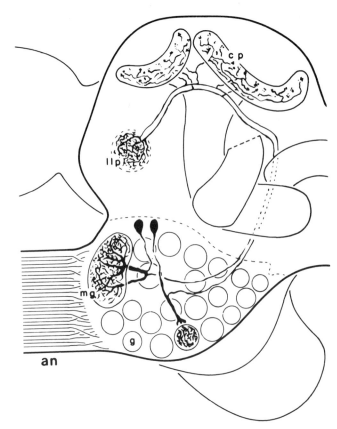

Figure 1. *Pheromone communication* in the silkmoth *Bombyx*. a. The two female moth pheromone components, *bombykol* and *bombykal*, each a specific stimulus to one of the two male odor receptor cells (see RC in f) in a hair sensillum. b. Schematized extracellular stimulus response: odor puff (top), generator potential (middle), impulses (bottom); recording as in f. c. Upwind path of the flightless male moth (♂) to its calling female (♀) or to pure bombykol (but not bombykal). d-h. Progressive enlargements of male antennal structures. d. Male with comb-shaped antennae, wingspan 40 mm. e. Section of antennal stem with branches and hair sensilla. f. Opened branch with hairs and receptor cells: (ST) sensillum trichodeum, (RC) receptor cell, (H) hemolymph, (RE) recording electrode, (IE) indifferent electrode. g. Section of a hair with pores (P), pore tubules, and two dendrites (D). h. Epi- and endocuticle with pores, pore tubules (PT), and a dendrite with microtubules.

Figure 2. Right half of the brain of a male cockroach in a frontohorizontal section plane. One output neuron of the deutocerebrum, sensitive to female sexual attractant, innervates the macroglomerulus (mg). a mechanosensitive + olfactory (fruit odor) output neuron innervates a regular glomerulus (g) more ventrally. Protocerebral projections are in the calyces (cp) of the ipsilateral corpus pedunculatum (mushroom body) and in the lateral protocerebral lobe (llp). Neurons were drawn by camera lucida from cobalt injected preparations; an, antennal nerve.

and cockroaches the endings of the receptor neurons that respond to the female attractant pheromone contact the interneuron endings in a specialized macroglomerular complex near the entrance of the antennal nerve into the brain. Still less is known of further processing of odor information in insect central nervous system (CNS), even with the rather simple, two- to five-component pheromone messages of moths. Receptor food odor discrimination and signal processing in cockroaches is similar to vertebrate function and is a promising example for a general understanding of olfaction.

Insects as model systems for study of odor receptive mechanisms are useful in many respects. In relative terms, no other olfactory system in the animal kingdom is better understood than that of the insects. But the tiny geometric dimension of the sensilla and of the brain require special topochemical and microphysiological methods to further analyze transduction and to understand CNS processing of a multitude of receptor messages. This, in fact, is the classical, still unsolved odor problem: How do we and the animals discriminate great numbers of odorants and their mixtures?

Further reading

Boeckh J, Ernst K-D (1983): Olfactory food and mate recognition. In: *Neuroethology and Behavioral Physiology*, Huber F, Markl H, eds. Berlin: Springer-Verlag

Kaissling KE (1985): Chemo-electrical transduction in insect olfactory receptors. *Ann Rev Neurosci 7*

Schneider D (1984): Insect olfaction: Our research endeavor. In: *Foundations of Sensory Science*, Dawson WW, Enoch JM, eds. Berlin: Springer-Verlag

Steinbrecht RA (1984): Chemo-, hygro-, and thermo-receptors. In: *Biology of the Integument, Vol 1, Invertebrates*, Bereiter-Hahn J, Matolsky AG, Richards KS, eds. Berlin: Springer-Verlag

The Olfactory Bulb

Gordon M. Shepherd

The olfactory bulb is the primary center for transmission of olfactory information to the vertebrate brain. Its functions are (1) to receive and process the information from the olfactory receptor neurons, (2) to send this information to different parts of the olfactory cortex in the forebrain, and (3) to provide for integration and modulation of the information by means of pathways from midbrain and forebrain centers.

The olfactory bulb is present among the most primitive vertebrates, and has been highly conserved throughout vertebrate evolution, being absent only in the cetaceans (whales, dolphins) which lack a sense of smell. Its homolog in higher invertebrates (e.g., arthropods) is the antennal lobe.

In most vertebrate species the olfactory bulb rests behind or upon the cribriform plate, which separates the cranial cavity from the nasal cavity. Bundles of olfactory nerves from the olfactory epithelium pass through openings in the plate to connect to the surface of the olfactory bulb. Closely associated with the main olfactory bulb is the accessory olfactory bulb. In most species this is a smaller structure embedded in the main bulb on the dorsal posterior surface. Its main function is to transmit information from the vomeronasal organ (Jacobson's organ) via the vomeronasal nerve.

Synaptic organization

The main neuronal components of the olfactory bulb are relatively constant throughout the vertebrate series, though they are most highly differentiated and distinct in reptiles, birds, and mammals. Most superficial is the olfactory nerve layer, where the olfactory nerves interweave. Next is a layer composed of spherical regions (50–200 μm across) called glomeruli, where the individual olfactory axons branch and terminate. Here they make synapses onto the dendrites of the relay neurons (mitral and tufted cells) (see Fig. 1).

The mitral cell bodies lie in a deeper layer; they are connected to the glomeruli by primary dendrites, and they have several secondary dendrites which branch and terminate in the intervening external plexiform layer (EPL). The mitral cell axon gives rise to one or several collateral branches before leaving the olfactory bulb to enter the lateral olfactory tract, from which it distributes terminal collaterals to several olfactory cortical regions. Tufted cells have traditionally been considered smaller versions of the mitral cells, though recent work has shown that the two types are distinct in terms of their genetic determinants, aspects of their physiology and neurochemistry, and their axonal collateral and projection pathways.

There are two main types of interneuron within the olfactory bulb. Situated around the glomeruli are the cell bodies of periglomular (PG) cells. These send their dendrites into the glomeruli, and their axons laterally within the interglomerular regions. The other type is the granule cell. The cell bodies are densely packed, and form the granule layer (GL) deep to the mitral cell body layer. They have a central dendritic process and a peripheral dendrite which ramifies within the EPL. The dendrites are heavily invested with spines. Granule cells lack axons and are therefore an anaxonal cell, similar in this respect to amacrine cells in the retina. A third type of interneuron is the short-axon cell; these are of several subtypes, scattered through the granule and glomerular layers.

The olfactory bulb receives a rich innervation of centrifugal fibers from the rest of the brain. These are of three main types: (1) fibers arising from the anterior olfactory nucleus and other olfactory cortical areas, which function as feedback loops for sensory processing; (2) fibers from the nucleus of the horizontal limb of the diagonal band (NHLDB) which are part of the basal cholinergic system innervating the forebrain; and (3) fibers from the midbrain, which are part of the noradrenergic and serotonergic systems innervating the forebrain. These fibers terminate at different levels within the olfactory bulb.

Electrophysiological experiments have shown that olfactory axons excite mitral/tufted and PG cells through their axodendritic synapses in the glomeruli. There is some evidence that PG cells act as inhibitory interneurons for the mitral cells, through both their dendrodendritic and axodendritic synapses (see Fig. 1). There is strong evidence that mitral/tufted cells excite granule cells, and that granule cells inhibit mitral/tufted cells, providing for self and lateral inhibition of the mitral/tufted cells. Short-axon cells may provide inhibitory control of the granule cells. Centrifugal fibers make synapses onto mitral/tufted and PG cells in the glomerular layer, granule spines in the EPL, and granule cell bodies and dendrites in the GL. A variety of evidence points to inhibitory effects of these fibers on mitral cells. Through these effects, the corresponding centers in the brain can set or modulate the excitability of the bulbar circuits, and thereby influence the significance of the incoming olfactory information for controlling behaviors such as feeding and mating.

Neurotransmitters and neuromodulators

The clear lamination and neuronal types have made the olfactory bulb an attractive subject for identification of neuroactive substances. The results of recent research are summarized in Figure 1.

The input pathways to the olfactory bulb are neurochemically distinct. The olfactory nerves are rich in the dipeptide carnosine; in addition, they contain a unique 20-kdal protein called olfactory marker protein. The centrifugal fibers from the locus coeruleus and raphe nucleus contain noradrenaline and serotonin, respectively; those from the NHLDB contain acetylcholine.

Within the olfactory bulb, the best characterized neuronal type is the granule cell, which is believed to use GABA

Figure 1. This schematic diagram summarizes the main neuronal elements, synaptic circuits, and neurotransmitter neuromodulator substances that have been identified in the vertebrate olfactory bulb. PG, periglomerular; GABA, gamma-aminobutryc acid; DA, dopamine; Enk, enkephalin; ACh, acetylcholine; 5HT, serotonin; LHRH, luteinizing hormone-releasing hormone; Glu, glutamate; Asp, aspartate; SP, substance P; NA, noradrenaline; SOM, somatostatin.

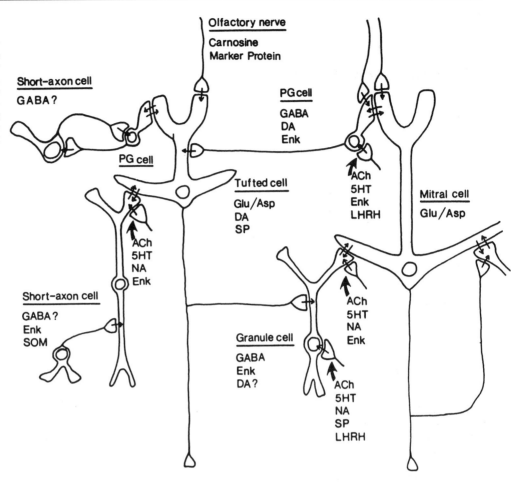

(gamma-aminobutyric acid) at its inhibitory dendrodendritic synapses onto the mitral cells. In the glomerular layer, some of the PG cell populations appear to be GABAergic, whereas some PG cells appear to be dopaminergic. These substances may be active at both the dendrodendritic and axodendritic types of synapses made by these cells.

The transmitters used by mitral and tufted cells are not known with confidence. The leading candidates are glutamate and aspartate. The olfactory bulb is particularly rich in neuromodulatory substances. It contains the highest concentration of taurine and thyroid-hormone releasing hormone in any brain region outside the hypothalamus. The distributions of enkephalin, substance P, somatostatin and luteinizing hormone-releasing hormone are indicated in Figure 1. Presumably these substances are important in modulating the activity of the olfactory bulb in relation to different behavioral states, such as occur in relation to mating and feeding.

Olfactory processing

Stimulation of the olfactory epithelium with pulses of odor gives rise to patterns of activity in olfactory bulb neurons that are temporally distinct and spatially distributed.

With regard to temporal patterns, a given mitral cell may respond, at threshold, with either excitation (a slow burst of impulses), or suppression (interruption of the resting discharge). With increasing odor concentration, the excitatory response changes to a brief burst followed by suppression. Mitral cells with these properties are believed to take part in the ensemble of cells that transmits specific information about

a particular type of odor molecule and encodes its concentration in the duration and intensity of the impulse discharge. Suppressive responses show suppression at all odor concentrations; cells with these properties are believed to be part of an extensive inhibitory surround.

Spatial patterns of activity within the olfactory bulb have been studied by several methods, including electrophysiological recordings and neuronal degeneration induced by chronic odor stimulation. The patterns have been seen most clearly using the 2-deoxyglucose (2DG) mapping technique. This has revealed domains of glomeruli activated by odor stimulation; significantly, the domains for different odors are overlapping but distinct. The domains show sharp boundaries, suggesting that they are due not to overall gradients of activity in the olfactory epithelium but rather to activity in distinct populations of olfactory neurons projecting to the glomeruli. These populations are beginning to be visualized by making small injections of horseradish peroxidase into the glomerular layer and tracing the retrograde transport into olfactory neurons in the olfactory epithelium. Within a glomerulus, a high-resolution 2DG analysis reveals a relatively uniform pattern of labeling, suggesting that a glomerulus, together with the olfactory neurons projecting to it, may be considered as a functional unit.

Development

The development of neurons and synapses in the olfactory bulb is closely associated with the development of the olfactory receptor neurons. These begin to be formed around day E10 in the rat. In single unit recordings, presumed receptor neurons can begin to respond to odor stimulation around E17. Mitral cells first appear at E11 in the rat. The first afferent axons reach the olfactory bulb by E15, and the first synapses are made within a day or two. Single unit responses of presumed mitral cells to odor stimulation have been recorded beginning at birth. The midbrain serotonin cells develop early (E11–15), but the arrival times of their axons in the olfactory bulb have not been determined. The intrinsic interneurons start to differentiate in the last few days of fetal life, and peak around P5; their transmitter (GABA) and many of the neuromodulatory substances make their appearances during this period.

This sequential appearance of types of neurons, synapses, and neuroactive substances lays the foundation for the sequential expression of function in different parts of the olfactory bulb during early life. Using the 2DG method, the earliest functional capacity has been detected in the rat in the accessory olfactory bulb, just before birth. Beginning immediately postnatal, the 2DG method reveals a focus of activity in a small, histologically distinct group of glomeruli termed the modified glomerular complex; it is believed that this complex, together with neighboring parts of the main olfactory bulb, mediates information about the odor cue on the mother rat's nipples that is essential for suckling by the rat pup to take place. At about 6 days, 2DG foci in the main glomerular sheet begin to be present in response to stimulation of the nose by various chemical odorants; these patterns appear to be consistent with those in the adult.

These studies suggest that olfactory function begins in fetal life. They indicate that the olfactory system consists of several subsystems, each developing according to its own timetable and mediating a function crucial for the organism at a particular stage of development.

Outputs to the forebrain

The output of the olfactory bulb is carried in the axons of mitral and tufted cells. These axons gather at the posterolateral part of the olfactory bulb to form the lateral olfactory tract, which courses over the basolateral surface of the forebrain. The axons send off numerous collaterals, which terminate in the superficial molecular layer of several distinct forebrain cortical regions. In rodents these include anterior olfactory nucleus, piriform cortex, olfactory tubercle, amygdala (cortical and medial nuclei), and transitional entorhinal cortex. Collectively, these constitute the olfactory cortex. There is also a medial olfactory tract that is particularly prominent in the fish. The accessory olfactory bulb has its own distinct projection to the cortical nucleus of the amygdala.

Through these direct connections to the forebrain, the olfactory bulb exerts its immediate control over feeding, mating, and a variety of related limbic behaviors. The wealth of feedback connections from the olfactory cortex and other limbic regions to the olfactory bulb by means of centrifugal fibers ensures that information processing by the olfactory bulb is always adjusted for the behavioral state of the animal.

Further reading

Halasz N, Shepherd GM (1983): Neurochemistry of the vertebrate olfactory bulb. *Neuroscience* 10:579–619

Macrides F, Davis BJ (1983): The olfactory bulb. In: *Chemical Neuroanatomy*, Emson PC ed. New York: Raven Press

Shepherd GM (1972): Synaptic organization of the mammalian olfactory bulb. *Physiol Rev* 52:864–917

Olfactory Discrimination

Maxwell M. Mozell

Although the term olfactory discrimination also includes the discrimination of different intensities of the same odor, it most generally refers to qualitative discriminations among different odors. There is a vast number of odor-producing chemicals. There is also a vast number of natural and artificial substances incorporating different sets of these chemicals, giving an even greater number of odorant mixtures. Furthermore, industries and laboratories continue to produce new chemicals and new substances, many of which are likely to give their own peculiar, heretofore unknown, olfactory experience.

Obviously, there is no practical way to test whether all the odor-producing chemicals and substances are truly discriminable from each other. However, our common experience of recognizing particular substances by their odors alone testifies to the tremendous number of odor discriminations which could likely be made. These discriminations can be quite subtle, as when we distinguish in a burning odor whether it is an electric circuit, an unextinguished cigarette, overheated brakes, or autumn leaves.

It has long been a question whether the olfactory system singles out individual odors in a mixture like the auditory system can single out the individual pitches of a chord, or, conversely whether the olfactory system accepts a mixture as a single odor totality analogous to the single color perceived when different wavelengths are mixed. We commonly smell many things that are mixtures of odorants but we tend not to experience them as simultaneous individual odors. Coffee, for instance, includes as many as 500 volatile compounds. On the other hand, perfume chemists can distinguish the various chemicals in a given fragrance, but this may in part be explained by their sophisticated knowledge of the mixture of odorants needed to produce different fragrances. From the available evidence, it appears that in some mixtures the individual odors can be perceived separately whereas in others they blend to give a unitary sensation.

A number of experiments have attempted to determine the perceptual or psychological dimensions along which odors are discriminated by asking subjects to judge how closely or how distantly the odors of a battery of odorants resemble each other. This defines an odor space, with each odorant occupying a place in the space relative to how closely its odor resembles the odor of each of the other odorants. The number of dimensions along which the odorants are apparently judged in order to define their relative places in the space can be estimated, but different studies have estimated different numbers (generally less than 5).

In order to name these dimensions, rather than simply enumerate them, one must evaluate from the order and spacing of the odorants along any given dimension what it is about their odors that is changing, and this has proved difficult. For instance, in a study suggesting two perceptual dimensions, one dimension is said to go from "a flowery, fruity, or gener- ally quite pleasant odor" to one which is "more spiritous or resinous." What term can be used to describe such a variation? The other dimension is said to be a variation in "sharpness or spiciness," which, though a more understandable name for a dimension, loses its credibility when the investigator describes the change in sharpness as a "slight tendency."

One experimentally determined dimension along which odors often appear to be judged is pleasantness. Everyday experience also indicates the clear effect that odors can produce; for many odors (e.g., fresh-baked bread, sweaty socks) it is difficult to remain indifferent to their presence. This apparent affective dimension of odors led one reviewer to suggest that the olfactory system is designed to answer two questions: What do I smell? Do I like it? Furthermore, it appears that the answer to the first question may be, in part, determined by the answer to the second question.

Several different mechanisms, each with creditable supporting evidence, have been proposed as possible bases for olfactory discrimination. One major contribution to our ability to discriminate between odorants likely derives from the multichannel input upon which the total smell sensation appears to depend. That is, in addition to the chemoreceptive endings of the olfactory nerve in the nose, there are chemoreceptive endings associated with the trigeminal, glossopharyngeal, and vagus nerves that line the respiratory tract at different levels (nose, pharynx, and larynx). These together define the accessory areas of olfaction, and these accessory areas, especially the trigeminal nerve, are credited with the pungent or irritating quality often sensed as part of an odor sensation. The magnitude of these inputs relative to that of the olfactory nerve differs for different chemicals, giving a possible basis for olfactory discrimination. The trigeminal input has often been suggested as underlying the pleasant-unpleasant dimension.

Over the years, the mechanism receiving most attention as being fundamental to olfactory discrimination has been the selective sensitivity of the individual receptor cells to different odorants. It was more or less expected, in analogy to some other sensory systems, that each receptor cell would be selectively sensitive to a particular set of chemicals. Cells responsive to the same set of chemicals could then be classified together as a basic cell type, and the differentiation among odorants would depend upon which of the basic cell types was being activated.

By and large, the many electrophysiological studies reporting the recorded responses of single receptor cells to batteries of odorants all reached the same conclusion, namely, each receptor cell did indeed show selective sensitivity in that it responds to a particular group of odorants, but, on the other hand, very few if any receptor cells respond to exactly the same set of chemicals. Therefore, the receptor cells cannot be classified into types although they are selectively sensitive. The mechanism for olfactory discrimination based upon selective sensitivity must, therefore, be somewhat more complicated than implied by the concept of receptor cell types. That is, each receptor cell apparently sees the world of odorants in its own peculiar way so that each odorant would establish a particular pattern of discharges from receptor cell to receptor cell across the entire population of receptor cells.

A number of these single receptor cell studies showed that to some degree odorants could be classified into groups according to the similarity of the cell-to-cell discharge patterns they elicited. This finding may seem paradoxical with the finding that the receptor cells cannot be classified into types by the odorants to which they respond, but there is at least one explanation. Perhaps there are several types of odorant receptor sites on each receptor cell, with most receptor cells having different sets and relative numbers of these receptor sites. Thus, two chemicals having some common property might excite the same receptor cells to similar degrees by virtue of one type of receptor site, but these same receptor cells might respond to other odorants very differently by virtue of their remaining, dissimilar sets of receptor sites.

By searching for common physicochemical properties among the odorants giving similar cell-to-cell discharge patterns, several investigators have been able to propose possible physicochemical dimensions along which the putative receptor sites could separate odorants. Among those so far proposed are absorption energy, molecular length, air/oil partition coefficient, oil/water partition coefficient, density, polarizability of the molecules, and van der Waals forces. It seems unlikely that there are separate receptor processes that measure any of these physicochemical properties alone. Rather these properties probably determine the access of odorant molecules to a variety of different receptor sites.

The possibility that there might be receptor sites having more selectivity than receptor cells has received support from at least two other lines of investigation. One of these exposes the mucosal sheet of receptor cells to reagents that bind irreversibly to particular protein groups and uses the electroolfactogram (EOG) to monitor the responsiveness of the receptor cells to different odorants. (The EOG is a slow odorant-produced potential change recorded from the mucosal surface that is believed to be the summed activity of many receptor cells in the vicinity of the recording electrode.) The premise to be tested is that the reagent, binding to some sites and not others, will specifically block EOGs for some odorants but not others. A further step in this approach is to determine how specifically EOG responses to different odorants can be protected from irreversible blockage by exposing the putative sites to each of the odorants before (or simultaneously with) the group-specific reagent. Although more conclusive interpretations could be reached if this blocking approach were coupled with recordings from single receptor cells rather than EOGs, there has been enough specificity in the results so far reported to continue entertaining the possibility of odorant-selective receptor sites.

Another line of investigation often interpreted as supporting specific receptor sites has been the search for specific anosmias. Some people are reported to have a much higher threshold than normal for a particular type of odor (as best demonstrated by a particular odorant) with otherwise normal thresholds for other odors. This is called specific anosmia, although, since sensitivity is reduced but not completely lost, specific hyposmia would be more fitting. The eight specific anosmias so far identified, together with the odorants best demonstrating them, are as follows: sweaty, isovaleric acid; spermous, 1-pyrroline; fishy, trimethylamine; malty, isobutyraldehyde; urinous, 5-androst-16-en-3-one; musky, ω-pentadecalactone; minty, 1-carvone; camphorous, 1,8-cineole. Assuming that specific anosmias can be further documented, one explanation for their existence, perhaps the most parsimonious, would be specific receptor sites. However, in spite of the indications, the existence of specific receptor sites is not yet certain. Several investigators propose that in the transduction process much of the cell membrane, rather than particular sites, can interact with the odorant molecules.

Different odorants establish different cell-to-cell activity patterns as determined by the different selective sensitivities of the various receptor cells. Several investigators have reported evidence suggesting that receptor cells particularly sensitive to the same odorant tend to cluster together in the same region of the mucosa, with different regions displaying different odorant sensitivities. That is, when different odorants are puffed directly onto various regions of the exposed olfactory mucosa, each odorant produces its largest EOG over a region that differs to some degree from those of the other odorants. This regional representation of odorants could be another basis for olfactory discrimination. It is known that there is at least some topographic projection of the mucosa into the central nervous system so that this spatial analysis of odorants at the mucosa level would be available for more central processing. Indeed, different odorants produce different topographic patterns of neural metabolic activity in the olfactory bulb as measured by 2-deoxyglucose uptake, but it is not yet clear that this reflects the regional selectivity of the mucosa.

Another possible mechanism underlying olfactory discrimination does not depend upon the selective odorant sensitivity of the receptors, but instead depends upon how the molecules of different odorants are spatially and temporally distributed to the receptors at their locations along the mucosal sheet. Electrophysiological recordings taken from branches of the olfactory nerve supplying earlier and later positions along the mucosal odorant flow path showed that, in a given sniff, the magnitude of the activity elicited at the later position compared to that at the earlier position ranged for different odorants from near equal to essentially nonexistent. The greater this spatial decrease in response, the longer was the time interval between the onset of the activity at the two positions. These findings could not be explained simply by differences in the odorant sensitivities of the receptors at the two positions since reversing the flow direction across the mucosa reversed the positions giving the larger and smaller responses. Instead, as suggested by using radioactively labelled odorants and by measuring the retention times of odorants flowing over a frog's olfactory mucosa, the explanation centered around the differential sorption of the molecules of different odorants by the mucosal sheet. The more strongly they are sorbed, the greater are their mucosal numbers early in the flow path, the fewer are their numbers later in the flow path, and the longer it takes them to travel particular distances along the flow path. It was argued that this process is analogous to the process occurring in gas chromatography and is sometimes termed the gas chromatographic model.

Thus, two possible mechanisms, regional sensitivities and molecular distributions, would use mucosal space to analyze odorants. Perhaps the strategy by which odorants are sniffed to and across the mucosa is important to olfactory discrimination, allowing an exploration of the olfactory stimulus analogous to the way the finger receptors explore tactile objects. The data in this regard is minimal, although one study observed that rats produce somewhat different sniffs when smelling different odorants.

Of the several mechanisms proposed at the mucosal level to underly olfactory discrimination, one may ultimately emerge as the one true mechanism. However, some or all of them may actually complement each other by giving, in their interactions, more response permutations for the encoding of odorants than any one mechanism alone.

Acknowledgments

Preparation of this manuscript was supported in part by NIH Grants NS03904 and NS19658.

Further reading

Engen T (1982): *The Perception of Odors*. New York: Academic Press

McBurney DH (1984): Taste and olfaction: Sensory discrimination. In: *Handbook of Physiology, vol III, sect 1. The Nervous System, pt 2*, Brookhart J, Mountcastle V, Darian-Smith I, Geiger SR, eds. Bethesda: American Physiological Society

Olfactory Psychophysics

William S. Cain

Chemists in the 19th century were the first to inquire about the smallest amount of airborne material detectable by the nose. As they suspected, the human sense of smell proved very sensitive to some materials, with reliable detection of a few parts per trillion parts of air, and relatively insensitive to others. In recent times, thresholds even an order of magnitude below parts per trillion have been measured. The green bell pepper odorant, 2-methoxy-3-isobutylpyrazine, with a threshold of 0.5 ppt, provides an example (see Table 1).

Threshold values, which now exist for about a thousand of the approximately half million odoriferous materials, span about 13 orders of magnitude. An absence of uniform methodology for threshold testing makes study-to-study comparability somewhat inexact, but some investigators have nevertheless used the body of existing data to search for physical and chemical determinants of sensitivity. The observation that threshold normally decreases with increasing chain length of molecules in aliphatic series suggested that solubility and perhaps molecular size play determining roles in sensitivity. Comparisons across series and among diverse odorants have continued to implicate those factors. Ability of an odorant to donate or accept protons and the existence of local sources of polarity within a molecule constitute solubility-relevant features that, in addition to molecular size, can portray the tendency of an odorant to leave the air, enter the mucous layer overlying the olfactory receptors, and pass through the mucus to the phospholipid membrane of the receptor. Solubility models leave some unexplained variance, but their relative success encourages the conclusion that mere migration of sufficient numbers of molecules to the receptor membrane per unit time largely determines threshold sensitivity.

The physicochemical and molecular properties that determine human olfactory sensitivity seem to determine animal sensitivity as well. Relative sensitivity is similar from human to dog to rat to frog, and so on. Absolute sensitivity may, however, differ considerably. Rats and dogs have sensitivity about three orders of magnitude greater than humans.

The sense of smell was traditionally reported as insensitive to small changes in odorant concentration. The measured Weber fraction (delta C/C, where C represents concentration and delta C represents a just-discriminable change in concentration) commonly equaled about $\frac{1}{3}$. This value, however, represents the net effect of all factors that can limit intensity resolution, including moment-to-moment fluctuations in the airborne concentration of the stimulus. It turned out that such fluctuations accounted in part for the high measured Weber fraction. With the fluctuations minimized or statistically discounted, the Weber fraction has fallen in the vicinity of $\frac{1}{10}$, which implies rather keen intensity discrimination. How much variation occurs across odorants awaits study.

The span of concentration necessary for perceived odor intensity to extend over the range from just detectable to very strong varies widely from one odorant to another. For odorants with very low thresholds, the span may equal upward of 10 orders of magnitude. For odorants with high thresholds, the span may equal only 2–3 orders. Exponents of psychophysical power functions fitted to data obtained from magnitude estimates of odor intensity range accordingly from the very low value of about 0.1 to about 0.7. Odorants with high thresholds and high exponents often exhibit pungency at concentrations not far above threshold. Such pungency comes about from stimulation of the trigeminal nerve and may influence the size of the exponent.

Adaptation has a potent influence on perceived odor intensity. A stimulus of constant concentration will wane in perceived magnitude exponentially over a few minutes and, if initially weak, may become undetectable. The rate at which the waning occurs seems rather similar across odorants. Psychophysical functions derived under conditions of adaptation suggest that olfaction, like vision, responds better to changes in stimulation than to steady stimulation. Dichorhinic stimulation under adaptation has revealed further that olfactory adaptation derives from both receptor events and central neural events.

Adaptation induced by one substance will alter sensitivity to some other substances, particularly those of similar structure and functionality. Often, this phenomenon of cross-adaptation proves quite asymmetric, i.e., prior inhalation of odorant A will depress sensitivity to odorant B much more than inhalation of B will depress sensitivity to A. Such asymmetries and unexpectedly strong cross-adaptation between some dissimilar odorants have so far foiled attempts to incorporate cross-adaptation data into any olfactory theory. Interestingly, cross-adaptation typically leaves the perceived quality of the test substance unaltered and induces no afterimages. In this respect

Table 1. Odor Detection Thresholds of Selected Compounds

Compound	Threshold (ppm, v/v)
Ethane	120,000
Methanol	500
Chloroform	65
Benzene	5.2
Camphor	0.17
Furfural	0.059
Isoamyl acetate	0.0071
5α-Androst-16-en-3-one	0.00019
2-Methoxy-3-isobutylpyrazine	0.00000054

Modified from Amoore JE (1982): Odor theory and odor classification. In: *Fragrance Chemistry: The Science of the Sense of Smell.* Theimer E, ed. New York: Academic Press.

it differs from cross-adaptation in taste and chromatic adaptation in vision.

Mixtures of odorants invariably smell weaker than the sum of the perceived intensities of their unmixed components. Mixtures of certain proportions will commonly smell even less intense than their stronger components smelled alone. Vector addition can satisfactorily describe the relation between the intensity of a mixture and that of its unmixed components for binary and ternary mixtures, but not for higher order mixtures. The parameter (angle between the vectors) estimated in the model shows no systematic variation with type of odorant. Mixtures of dissimilar odorants therefore exhibit about the same degree of perceptual additivity as mixtures of similar odorants. This outcome would suggest that odor masking, i.e., suppression of the perceived intensity of a weaker by a stronger component, will also follow an invariant rule across odor qualities.

Perceived odor quality has generally eluded satisfactory quantification. Direct estimates of similarity lack the precision for resolution of small, yet distinctive differences. Conversely, open-ended and often metaphoric descriptions of the sort made by perfumers resist quantification by their nature. Attribute scaling, which entails rating the applicability of a large pool of nonredundant descriptors (more than a hundred for general purposes) to an odorant, seems to offer the necessary precision and stability when obtained from a group of about 50 observers. Attributes have typically been chosen for this purpose on strictly empirical grounds, without imputation of primacy. Once-popular notions that all odor experience could be created from mixtures of a few primaries or that all odor experience would prove derivable from some fundamental attributes have largely disappeared.

Although odor quality has proved somewhat difficult to measure, odor identification has proved useful for rapid clinical evaluation of olfaction. Standardized multiple choice tests with from 10 to 40 everyday odors offer data sufficiently precise to characterize patients as normosmic (normal smell); mildly, moderately, or severely hyposmic; or anosmic. Growing medical interest in disorders of the sense of smell now provides a clear purpose for advances in diagnostic olfactory psychophysics.

Further reading

Carterette EC, Friedman MP, eds (1978): *Handbook of Perception, Vol 6A, Tasting and Smelling.* New York: Academic Press

Engen T (1982): *The Perception of Odors.* New York: Academic Press

Meiselman HL, Rivlin RS, eds (1985): *Clinical Measurement of Taste and Smell.* Boston: Macmillan

Olfactory System, Turnover and Regeneration

P.P.C. Graziadei and G.A. Monti Graziadei

The olfactory sensory epithelium lines portions of the nasal cavity, and in the adult animal three cell types, which are organized in columnar units, contribute to its formation: neurons, basal cells, and supporting cells (Fig. 1). The axons of the olfactory sensory neurons collect in fascicles of increasing diameter in the lamina propria mucosae from where they run centripetally toward the olfactory bulb. In their course they either assemble in a unique olfactory nerve (the olfactory nerve of lower vertebrates) or maintain their arrangement in several discrete bundles (the fila olfactoria of mammals) that penetrate into the cranial cavity through the lamina cribrosa. The olfactory axons form an intricate nerve plexus on the surface of the homolateral olfactory bulb before terminating in globose structures, the olfactory glomeruli, where they synapse with the dendrites of large cortical neurons, the mitral cells. The olfactory sensory pathway is, consequently, the only one that directly extends from the periphery to a cortical area without entering a commissure and without a synaptic interruption.

The olfactory neurons of vertebrates, from fish to mammals, are continuously replaced (turnover) during adult life, and this phenomenon probably occurs in humans as well. The basal cells, located close to the basal lamina, have recently been identified as the sole precursors of mature neurons, and morphological and autoradiographic studies have helped to identify the stages of maturation of these neurons (Fig. 2). According to the number of immature neuronal elements in different areas of the olfactory neuroepithelium, two spatially distinct zones can be found interspersed across the sensory sheet. They have been described as quiescent and active. The quiescent zones are characterized by a prevalence of mature neurons and are presumably the ones functionally important. The active zones have an abundance of developing elements, and it is here that the process of neuron formation occurs (Fig. 3).

In the last decade several experimental paradigms have elucidated the dynamics of this unusual neuron and have recognized some of its fundamental properties. Severance of the olfactory sensory axons induces irreversible degeneration of the sensory perikarya. The degenerative process, contrary to what occurs for all other neurons of the nervous system, is followed by an outburst of mitotic activity in the basal cells that subsequently differentiate and mature into sensory neurons. Even more interestingly, their axons are capable of elongating and establishing anatomical and functional connections with their proper target (the mitral cells), as demonstrated by complementary anatomical, electrophysiological, and behavioral observations. This is the only sensory pathway that after being totally interrupted can, even in mammals, be reconstituted both anatomically and functionally. These experiments suggest that, in the adult animal, guidance mechanisms allow regrowing olfactory axons to reestablish an adequate pattern of projections between the peripheral sensory sheet and the glomerular layer of the olfactory bulb. These experiments indicate that this

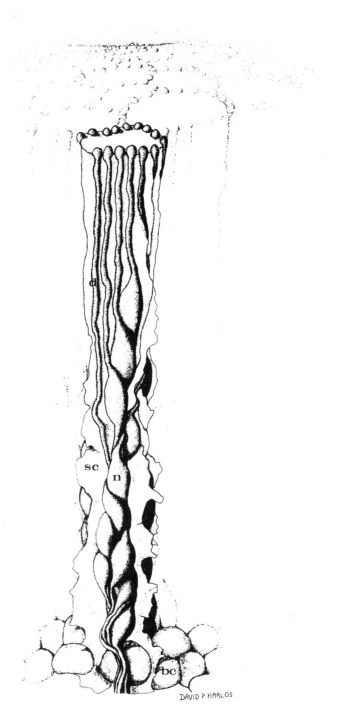

DAVID P HARLOS

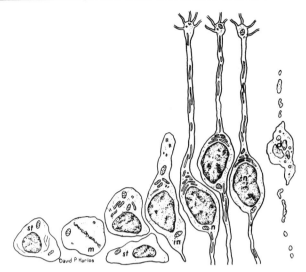

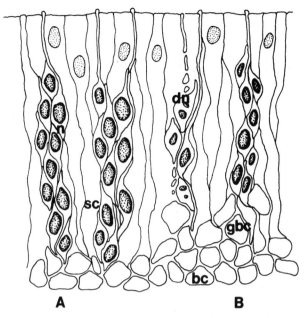

Figure 2. The stages of differentiation and maturation of the olfactory neurons as observed in the normal olfactory mucosa. Staminal cells (st) frequently divide (m) and differentiate into immature neurons (in). In n are indicated the mature neurons and in d the debris of a degenerated neuron. From *Neuronal Plasticity*, Cotman CW, ed. New York: Raven Press, 1978.

Figure 3. Diagram of the olfactory epithelium with a quiescent (A) and active (B) zone. The basal immature elements (bc and gbc) are numerous in B. Between quiescent and active zones a portion of the diagram shows the intermediate process of neuron degeneration (dn). (sc) supporting cells. Reprinted with permission from *Journal of Neurocytology* (1979). 8: 1–18. London: Chapman and Hall Ltd.

portion of the telencephalic cortex (the bulbar cortex) remains accessible to the penetration of new axons and that adequate synaptic sites are available to assure the return of appropriate functional contacts.

When partial ablation of the olfactory bulb is performed concurrently with interruption of the sensory pathway, the axons regrow and reinnervate the olfactory bulb, but the topographical organization of the glomeruli is changed. In fact, new glomeruli form in the deep layers of the bulb such as the external plexiform layer and the granule layer. The ectopic arrangement of glomeruli is accompanied by reorientation of the dendrites of the mitral cells and by the establishment of new synaptic contacts. Preliminary behavioral results indicate that odor perception is maintained, even in these abnormal anatomical conditions, and it is not known what level of function is supported by such morphological changes.

After removal of the olfactory bulb, the sensory neuron population continues to turn over. Olfactory axons can readily penetrate the forebrain regions, which have been exposed by the bulbar ablation, and here they form the characteristic glomerular structures that serve to establish synaptic contacts with dendrites of regional neurons. These experiments indicate that the olfactory neurons do not require a specific target for expressing their characteristic terminal structures, the olfactory glomeruli.

Experiments in amphibians have confirmed and extended the results obtained in mammals. The olfactory organ can be transplanted, at early embryonic stages, to other parts of the head or body without impairing its development. In fact, an olfactory nerve grows from the ectopically located organ, and when it reaches the central nervous system, it induces profound morphological modifications. For example, in *Xenopus*, substitution of the eye cup with the olfactory placode inhibits eye regrowth. From the mutilated optic stalk a new lobe originates, and the olfactory axons, which penetrate into this new lobe, form glomerular structures similar to the ones observed at the cranial end.

Further reading

Graziadei PPC, Monti Graziadei GA (1979): Neurogenesis and neuron regeneration in the olfactory system of mammals. *J Neurocytol* 8:1–18

Ottoson D (1983): *Physiology of the Nervous System*, Chap. 29: Olfaction, New York: Oxford University Press, pp. 429–447

◁ **Figure 1.** Schematic illustration of a portion of a columnar unit showing the columnar arrangement of the mature neurons (n) and the position of their dendrites (d) around the supporting cells (sc). (bc) basal cells. Reprinted with permission from *Journal of Neurocytology*, (1979) 8:1–18. London: Chapman and Hall Ltd.

Pain, Animal

Howard H. Erickson

Pain is a complex physiological phenomenon; it is hard to define satisfactorily in human beings, and it is extremely difficult to recognize and interpret in animals. Scientific knowledge of pain perception in animals has been obtained by drawing analogies based on comparative anatomy, physiology, and pathology, and by inference based on subjective responses to pain experienced by humans. Debate continues about whether animals of different species perceive pain similarly and whether any species perceives pain the same way humans do. Our knowledge of the scientific basis of the mechanisms of pain in animals, however, has advanced substantially in the last two decades.

Nociceptors, or pain receptors, are widespread in the skin and tissues of animals and are often considered free nerve endings. The structural and physiological basis of receptors is, however, an elusive area of sensory research. Chemical mediation of nociceptor excitation and its role in neurogenic inflammatory responses may be a key to understanding many of the peripheral phenomena related to pain.

Neurophysiological mechanisms of nociception operate at the segmental level of the spinal cord in domestic animals. There are four main aspects at the segmental level: (1) the sensory input, (2) the neurons of the dorsal horn, (3) the influence of descending control mechanisms, and (4) the output from the segmental level. Although a specialized pain signaling mechanism seems to exist within the spinal cord, no simple mechanism has been identified.

The responses of animals to noxious stimuli resemble those of humans and include both sensory and motor reactions. Nociceptive signals are used by various species to trigger both sensory-discriminative and motivational-affective processes. The relative importance of ascending spinal cord pathways that transmit nociceptive information in animals varies with the species.

Endogenous neural systems in the brain stem and forebrain, including both opioid and nonopioid mechanisms, modulate the central transmission of nociceptive signals in animals. The complete extent of brain areas contributing to descending modulation of spinal nociceptive transmission is not yet known, but widespread regions of the brain stem and forebrain are involved. Descending inhibition of spinal dorsal horn neurons may be an important mechanism underlying the analgesic effect of central stimulation. The level of activity in each descending control system may be subject to a complex array of inputs coming in from other brain areas.

Stimulation analgesia and endorphins are some of the liveliest areas of pain research. Central nervous system substrates, the opiate peptides, enkephalins and endorphins, are involved in the inhibition of pain, in addition to a nonopioid mechanism which has a neurohumoral basis involving adrenal enkephalins and the pituitary-adrenal axis. This intrinsic analgesia system in the brain stem produces an endogenous substrate that results in important adaptive behavior.

There are a variety of stimuli (superficial, mechanical, thermal, chemical, and visceral) which can evoke pain and behavioral changes in animals. The perceptual experience can change dramatically with variation of the intensity or characteristics of stimulation. Although nonhuman species cannot make qualitative distinctions between pain and sensations by verbal reports, it is feasible to assess the intensity of elicited pain by observing the intensity of behavioral reactions that eliminate painful stimulation. Comparison of human and animal data shows that animals begin to escape stimulation at about the same intensity that human subjects first report pain, confirming the validity of the escape paradigm.

Stimuli are considered noxious to animals if they produce pain or tissue damage in humans or escape behavior in animals. The relief of surgical pain is primary justification for the existence of anesthesia. It is essential to properly assess pain and anesthesia during surgical procedures in experimental animals. Unfortunately, an objective index does not exist that specifically and reliably correlates with pain perception. There is no unequivocal best anesthetic technique for all animals, investigators, and procedures, but there are steady-state descriptors that can be used to judge the level of anesthesia. Depth of anesthesia is estimated from changes in observed physiological variables, which include heart rate, arterial pressure, arterial or end-tidal carbon dioxide, eye reflexes, and muscle tone.

Exposure to noxious or potentially lethal forces is virtually unavoidable for most animals. The ability to detect stimuli, such as predators, and to take action to minimize their effects represents a major selection pressure in animal phylesis. Natural selection based on nociception might have shaped the emergence of present-day species and their behavioral repertoires. Three selection pressures can be identified as potentially relevant to the phylogenetic evolution of pain: (1) the exigencies imposed by overt trauma, (2) the demands imposed by the threat of injury, and (3) the social consequences of pain experience and expression.

Considerable differences in the disposition of analgesic drugs are encountered among common domesticated animals. Knowledge of these differences is crucial to the rational use of drugs in therapy and to the alleviation of pain. These considerations affect the selection of drugs, dosage, and routes of administration for alleviating pain in domestic animals. Differences in absorption, distribution, storage, excretion, and biotransformation are responsible for most of the variation in drug responses encountered among species.

There is a large species variation in the absorption and biotransformation of drugs used to alleviate pain in animals. Most of the pharmacokinetic differences observed among species are due to varying rates or different pathways for biotransformation. In some species the duration of action is too short or there is extreme sensitivity to the drug. For example, the plasma half-life of salicylate is 38, 8, 6, 1, and 0.8 hours in cats, dogs, swine, horses, and goats, respectively. Since dogs and cats metabolize 85–90% of a dose of meperidine within an hour after administration, this drug is not very useful for the management of pain compared with morphine.

In general, carnivores conjugate foreign chemicals slowly, herbivores rapidly, and omnivores with intermediate rapidity. The mechanisms of action of some analgesics are also different in different species. Domestic animals seem to oxidize various drugs rapidly, in contrast to humans. Several specific differences are known in drug-metabolizing enzymes. In general, most drugs are metabolized more rapidly by laboratory rodents and more slowly by humans than by domestic animals.

Of the domesticated species of animals, horses are probably most refractory to the effects of analgesic agents. A great need exists for effective agents that produce a sedative and analgesic effect in this species. An equine pain model has recently been developed in which three kinds of pain (superficial, deep, and visceral stimuli) can be produced and analgesic drugs objectively evaluated.

In different breeds of dogs, variability in response to pain occurs not only within and among species but from individual to individual. Apparently, it is the tolerance to pain that is highly variable even though pain thresholds are remarkably similar among animal species. Pain itself is a sign of injury and aids in healing by restricting activity that would interfere with healing. Only after determining if the animal can be made more comfortable by the use of pain-relieving drugs, and that these agents will not injure the animal or allow for potentiation of an existing injury, should a pain reliever be selected.

These examples illustrate some problems involved in the alleviation of pain in different species of animals, and why it is unwise to extrapolate information concerning drugs from one species to another. As the mechanisms of pain are better understood, the humane treatment and alleviation of pain in animals can be placed on a much firmer scientific basis.

Further reading

Erickson HH, Kitchell RL (1984): Pain perception and alleviation in animals. *Fed Proc* 43:1307–1312

Kitchell RL, Erickson HH, Carstens EA, Davis LE (1983): *Animal Pain: Perception and Alleviation*. Bethesda: American Physiological Society

Pain, Chemical Transmitter Concepts

Lars Terenius

The word pain covers phenomena ranging from the acute sensation leading to a nocifensive response to long-term or chronic suffering, a disease state. When cross-species comparisons are made, the general evolutionary level of the species involved must be considered. Since pain is such a loose term, better definitions are needed. These may include descriptions of the chemical substrates and neuronal transmitters involved in pain and pain modulation, which are likely targets for intervention by pharmacotherapy.

A main category of pain is the acute sensation, the sharp easily observable component, which has its origin in cutaneous tissue and is conducted via A-delta-fiber afferents. A second category, slow pain or ache, is elicited by noxious, i.e., tissue-damaging stimulation of skin and deeper tissues, muscle, joints, or viscera and is mainly conducted via nonmyelinated C-fiber afferents. Both categories of pain can be ascribed to a stimulus of strong intensity. However, in certain kinds of nerve damage normally nonnoxious stimuli may cause the manifestation of pain, a phenomenon called allodynia. Finally, pain may also be a complaint in the absence of apparent noxious stimuli or evidence of nerve damage. It may then be better explained in psychological terms, so-called psychogenic pain. Apparently there is no direct correspondence between the stimulus (injury) and pain experiences.

Primary afferents

It is presently unknown what chemical mechanisms are involved in the activation of A-delta-fibers or what the transmitter of these fibers can be. The motor reflex to a nociceptive stimulus seems resistant to specific pharmacologic intervention. Recent studies suggest, however, that the neuropeptide cholecystokinin (CCK), present in interneurons in the reflex arc, may modulate the response. Several groups have reported intrathecally given CCK to have apparent morphine antagonistic properties. C-fiber afferents have been much better characterized with regard to mechanisms of activation, transmitter content, and anatomical distribution. Most C-fibers have polymodal nociceptors; i.e., the peripheral nerve endings respond to strong mechanical and chemical stimulation besides being heat sensitive. Different groups of fibers may show different thresholds with regard to these stimuli, depending on whether they originate in skin, deep muscle, joints, or viscera. Since most experimental work has been done with afferent units from skin, information is needed to determine how representative these are for fibers of other origins.

Several endogenous low-molecular-weight substances elicit pain by activating polymodal nociceptors. There are several groups of such agents: peptides (bradykinin), amines (serotonin, histamine), and arachidonic acid derivatives (prostaglandins, prostacyclin). All these agents are produced by injured tissue from precursors and by mechanisms that are ubiquitously available: bradykinin from kininogen, which is activated by kallikrein, an enzyme of the blood-clotting cascade; amines released from mast cells; and the arachidonic acid derivatives formed intracellularly in virtually every cell type. Bradykinin is probably the most potent algogenic (pain-producing) irritant, potentiated by arachidonic acid derivatives, which by themselves apparently have very little algogenic activity. Polymodal nociceptors seem to have separate receptor mechanisms for the peptides, amines, and arachidonic acid derivatives. A population of nociceptors may in fact become desensitized to one group of algogenic agents while maintaining full response to others. Nociceptors also respond to more nonspecific exogenous irritants such as mineral acid. Aspirin and its congeners block the enzyme cyclooxygenase in the arachidonic acid prostaglandin biosynthetic pathway.

Several neuropeptides have been identified in C-fiber afferents by immunohistochemical analysis; 30–50% contain substance P, less than 30% somatostatin or vasoactive intestinal polypeptide (VIP). C-fiber afferents from tooth pulp are almost exclusively nociceptive and contain mainly substance P. Visceral C-fiber afferents may have a higher proportion of VIP fibers than cutaneous afferent C-fibers. Substance-P-containing afferent fibers terminate in lamina 1 and the underlying substantia gelatinosa of the spinal cord; VIP fibers, mainly in lamina 1. A considerable number of other neuropeptides have also been identified in thin, nonmyelinated afferent fibers including calcitonin gene-related peptide (CGRP), and neuropeptide tyrosine (NPY). Partly for historical reasons there is by far more information on the role of substance P in primary afferents than for the others.

Substance P is synthesized in the perikarya of the dorsal root ganglia. To act as a conveyor of sensory information, it is transported by axonal flow centrally to terminals in the spinal cord. Several experimental findings support the role of substance P as a transmitter in these nerves: high-threshold electrical stimulation of afferent nerves releases substance P in spinal fluid; microinjection of substance P into spinal cord produces electrophysiological responses similar to those caused by noxious stimuli; substance P given intrathecally elicits behavioral responses suggesting irritation; and administration of antiserum to substance P, or peptides with substance P antagonistic activity, produces hypoalgesic responses. The pungent principle of red pepper, capsaicin, depletes substance P from primary afferents and reduces pain sensitivity in experimental animals. It has been suggested that substance P may lower the threshold for postsynaptic excitation by another pain transmitter. The afferent C-fiber may thus provide an example of the coexistence of a conventional transmitter substance and a neuropeptide, the former being the primary neurotransmitter. Supporting this possibility is the relatively slow on- and offset of action of microiontophoresed substance P on spinal nerve cells. Candidates for conventional transmitters in C-fibers are excitatory amino acids (glutamate or aspartate).

The major part (maybe as much as 90%) of substance P synthesized in the dorsal root ganglion cells is transported in a centrifugal direction. There is evidence to suggest that this substance P is the mediator of the axon reflex, which is responsible for the slowly spreading flare (neurogenic inflammation) that is part of the nocifensor system. Substance P released from the peripheral nerve ending causes vasodilation, release of histamine, and an increase in vascular permeability. Histamine adds to the vascular effects of substance P and also stimulates neighboring C-fiber terminals.

Local circuitry of the spinal cord

The termination area of the C-fiber afferents in the dorsal horn of the spinal cord has a complicated anatomy. Available for synaptic modulation of primary afferent transmission are two opioid peptide systems, mainly of local nature, and neurons, some of which are serotonergic, which descend from the brain stem.

Some biochemical characteristics of the opioid peptide systems are summarized in Table 1. Like other neuropeptides, their biosynthetic origin is a prohormone; each proenkephalin molecule can give rise to seven enkephalins of slightly different structures. In an analogous manner, prodynorphin can give rise to three different opioid molecules. The two systems have different biochemical and genetic origins and are seldom simultaneously expressed in one neuron.

Several layers of the dorsal horn are richly innervated by enkephalin fibers. Most of these fibers are interneurons. The older literature suggested termination of enkephalin neurons on the afferent substance P neuron terminals. This has not been verified by histological techniques. However, there is considerable topological overlap between substance P and enkephalin terminals. Nonsynaptic interaction may be of importance, and opiate receptors on substance P terminals have been proposed to suppress substance P release. Dynorphin nerve terminals have a more discrete distribution in the dorsal cord, with major distribution in laminas 1 and 4, areas of particularly dense C-fiber afferent innervation. Dynorphins are also mainly in interneurons, although there is a suggestion of dynorphin afferent fibers.

As indicated in Table 1, the enkephalin and dynorphin families differ in receptor preferences. The affinity of these peptides for the classic opiate receptor (mu-morphine receptor) is about 10–50 times less than for delta and kappa receptors, respectively. Biochemical receptor mapping has indicated rich occurrence of kappa receptors in the spinal cord, especially in humans. The involvement of dynorphin (and kappa receptors) in pain modulation may be exclusively at the spinal level; dynorphin given intrathecally is a potent analgesic, whereas it has not been shown to be active after microinjection into the brain. Experimental drugs with kappa receptor profile seem particularly efficient against chemically induced pain, whereas

mu receptor agonists are more efficient against heat-induced pain.

Serotonergic neurons with cell bodies in the raphe nuclei may directly interfere with transmission of the C-fiber afferent neurons or may project to enkephalin interneurons, which act secondarily. The importance of this pathway has been verified experimentally and clinically. Electrical stimulation of the periaqueductal gray matter in the midbrain produces powerful pain relief partly through these systems. Drugs blocking serotonin reuptake mechanisms, essentially the antidepressants, potentiate morphine analgesia.

Supraspinal systems

The nature of the transmitters in second-order ascending neurons is not known. Axon collaterals from the ascending pathways have rich contacts with other characterized neuronal systems, for instance, enkephalin interneurons in brain stem, midbrain, and thalamic areas. The third opioid peptide system, generating beta-endorphin, has a limited distribution in the central nervous system (CNS). However, the potency and stability of this peptide in CNS may contribute to its importance. Beta-endorphin fibers project to areas close to the third ventricle. Beta-endorphin may be secreted in neurohormonal fashion into cerebrospinal fluid and thus reach midbrain areas including the periaqueductal gray matter, an area responsive to microinjected opiates. This peptide is relatively unselective in receptor preference and has significant affinity for mu-type opiate receptors.

Microinjection of opiates into the nucleus raphe magnus and adjacent areas in the medulla causes powerful analgesia. Opiates microinjected in this area and in dorsal spinal cord act in synergy. Both midbrain and medulla areas have rich enkephalin innervation. These sites are probably highly significant for the analgesic action of opiates.

Enkephalin fibers are richly represented in limbic areas. Biochemical studies and autoradiography have demonstrated the differential distribution of the various opioid receptors. The mu receptor ligands accumulate in thalamic structures and deeper layers of cerebral cortex, possible targets for the action of opiates on the affective components of pain. Delta receptors are common in limbic areas.

Several neuropeptides produce analgesia if microinjected into cerebral ventricles or specific brain areas such as periaqueductal gray. Paradoxically this is also the case with substance P, which may act indirectly by provoking release of enkephalin.

Pain, stress, analgesia, and placebo

Acute pain impels a nocifensive response, a reaction of obvious survival value. Enduring pain could, however, interfere with the coping reaction. Several studies suggest that stress is a natural trigger of pain suppression. Experimental studies have been carried out in rodents with defined parameters of stress, frequently induced by inescapable foot shock of varying intensity, frequency, and duration. The analgesia observed can be reversed by naloxone, indicating a possible opioid link. The opioid analgesia is blocked by removal of the adrenal medulla, which does not affect nonopioid analgesia. The adrenal medulla is a rich source of peptides of the enkephalin family which are postulated mediators of the response. Both kinds of analgesia are blocked by surgical lesions of the spinal dorsolateral funiculus.

Strong enduring pain in humans increases blood flow and utilization of glucose in most brain areas, strongly emphasizing

Table 1. The Three Groups of Opioid Peptides and Their Receptor Preference

Opioid Peptide	Number of Amino Acids	Receptor Preference
Enkephalins[a]	5–8	delta
Dynorphins[b]	10–17	kappa
Beta-endorphin	31	mu and delta

[a] Leu-enkephalin, Met-enkephalin, Met-enkephalin-Arg[6]Phe[7], Met-enkephalin-Arg[6]Gly[7]Leu[8].
[b] Dynorphin A, dynorphin B, alpha-neoendorphin.

the generalized effects on CNS function. Emotional stress may also reinforce pain by sustaining the causative pathological processes. A classic example is pain induced by ischemia and amplified by stress-induced release of vasoconstrictors, notably epinephrine. To what extent simultaneous release of enkephalin peptides can modulate this reaction is unknown.

Techniques such as acupuncture and transcutaneous nerve stimulation (TNS) are becoming popular as clinical methods for pain relief. It is likely that they activate the same substrates as stress. Interestingly, TNS with high intensity and low frequency (about 1 Hz) produces pain relief that is at least partly blocked by naloxone, whereas TNS of low intensity and higher frequency (100–200 Hz) produces pain relief, which is generally not reversible by naloxone. Acupuncture in experimental animals and in humans has been found to affect activity in several CNS neurotransmitter systems including opioid, gamma-aminobutyric acid (GABA), norepinephrine, and serotonin. Elucidation of the chemical mechanisms of nonopioid stress analgesia and analgesia produced by TNS or acupuncture will ultimately increase the possibilities of using endogenous mechanisms of pain control for therapeutic purposes.

Behavioral techniques for pain relief, such as hypnosis, relaxation, or biofeedback are unaffected by concurrent naloxone administration. Expectation of pain relief in a clinical setting is by itself a strong modifier of pain. Recent literature on the mechanisms of this so-called placebo response is partly contradictory. Opioid mechanisms may or may not be involved. The design of such studies is critical and complicated. For instance, "the hidden injection" as placebo control, is hard to give unnoticed.

Pathologic, clinical pain

The many faces of clinical pain have already been alluded to. Several kinds of pain develop into chronic pain states and, by definition, are therapeutic failures. Chronic pain caused by a pathologic process, constant or progressing as in terminal cancer, obeys mechanisms comparable to those in acute pain. That pain is not eliminated is due to the inherent ineffectiveness and actual or potential side effects of common analgesic drugs. Neurogenic pain, which is secondary to lesions in peripheral nociceptive or central neurons, is commonly resistant to usual pharmacotherapies. This suggests a mechanistic difference from common nociceptive pain. Several lines of evidence suggest that this clinical condition is characterized by deficient pain modulation rather than excessive influx of information from primary nociceptive afferents. Chronic neurogenic pain is more frequently responsive to acupuncture and TNS than other kinds of pain. In comparison with other groups of patients, levels of opioid peptides in cerebrospinal fluid are regularly lower in these patients. Pain is a frequent complaint in patients with psychiatric disorder; pain is sometimes the main syndrome even if there is no objective organic cause. Such patients have a tendency for elevated opioid peptides in cerebrospinal fluid. These observations and others emphasize the importance of an appropriate taxonomy of chronic pain.

Further reading

Han JS, Terenius L (1982): Neurochemical basis of acupuncture analgesia. *Annu Rev Pharmacol Toxicol* 22:193–220

Salt TE, Hill RG (1983): Neurotransmitter candidates of somatosensory primary afferent fibers. *Neuroscience* 10:1083–1103

Terenius L (1981): Endorphins and pain. *Front Horm Res* 8:162–177

Terman GW, Shavit Y, Lewis JW, Cannon JT, Liebeskind JC (1984): Intrinsic mechanisms of pain inhibition: activation by stress. *Science* 226:1270–1277

Pain, Intracerebral Stimulation as Treatment

Björn Meyerson

Electrical stimulation via electrodes implanted stereotactically in deep-seated regions of the brain as treatment of chronic and severe pain has been in use since the mid-1970s. The surgical procedure of implanting electrodes is comparatively simple and carries little risk of complications, but this approach to the management of pain is practiced in only a few centers and on very strict indications. Intracerebral stimulation is of theoretical interest because of the insights it provides into the physiological mechanisms involved in the endogenous control of pain.

Intracerebral stimulation for pain may be applied in two different target regions: the specific sensory nuclei of the thalamus and the periventricular gray matter (close to the posterior part of the wall of the third ventricle). The pain relief resulting from stimulation in either of these regions is dependent on different physiological mechanisms corresponding to two different types of pain: nociceptive and neurogenic. The former is interpreted as being caused by the activation of specialized peripheral nociceptors. Neurogenic pain is a sensation perceived as painful but caused by injury to the nervous system. Neurogenic pain is usually resistant to peripherally and centrally acting analgesics, whereas nociceptive pain may be effectively relieved by these agents. Sensory thalamic stimulation preferentially relieves neurogenic pain; periventricular stimulation, nociceptive pain.

Intracerebral stimulation for pain illustrates how data from basic science can be transferred from the laboratory to the clinic and directly applied for therapeutic purposes. This form of treatment is closely linked to two major breakthroughs in modern pain research. It is well known that both transcutaneous nerve stimulation and stimulation of the dorsal columns of the spinal cord, which are now widely used in treatment of pain, evolved from the introduction of the gate-control theory. Though disputed in its details, the essentials of this theory, which postulates that impulses signaling pain may be profoundly modified by concomitant activity in large afferent fibers, have had an enormous impact on pain research. Although much experimental data are available on the intricate organization of the pain-transmitting relay in the dorsal horn of the spinal cord, it has been difficult to induce in animals effects corresponding to the long-lasting inhibition of chronic pain in humans following afferent or spinal stimulation. In some neurogenic pain associated with deafferentation, the neuronal substrate for transcutaneous or dorsal column stimulation is insufficient; typical examples are pain in extensive spinal cord lesions, avulsion of the brachial plexus, and anesthesia dolorosa. It is logical to consider the possibility of relieving such pain with stimulation of the supraspinal portions of the lemniscal system rostrally to the relay in the dorsal column nuclei, i.e., the specific sensory thalamus or the thalamocortical sensory projection in the internal capsule. Stimulation in these regions evokes paresthesias which must be felt in the painful area in order to produce pain relief. Generally, about thirty minutes of stimulation may significantly reduce the pain for 3 to 6 hours. This type of stimulation has been found to be particularly useful for posttherpetic facial pain. In other pain conditions the outcome is unpredictable and beneficial effects often do not last. Virtually nothing is known about the physiological mechanisms responsible for the pain blocking effect of supraspinal lemniscal stimulation. It has been hypothesized that supraspinal gating mechanisms similar to those in spinal modulation of nociception after transcutaneous nerve stimulation are responsible. However, experimental data are lacking. There is some circumstantial evidence that stimulation in sensory thalamus activates dopaminergic, and perhaps serotonergic, systems, but not endogenous opioides since the pain relieving effect cannot be reversed by naxolone.

The use of the periventricular gray area as a target region for intracerebral stimulation of pain is directly based on the observation by Reynolds in 1967 of stimulation-produced analgesia. He discovered an endogenous pain controlling system in the core of the brain stem, extending rostrally to the posterior part of the diencephalon, the activation of which may inhibit pain transmission at the first synaptic relay in the dorsal horn of the spinal cord. The ensuing demonstration of a close relationship between stimulation-produced analgesia and the antinociceptive mode of action of opiates initiated intensive research. A host of data has accumulated on the involvement in stimulation-produced analgesia of neurotransmitters or neuromodulators with morphine-like properties, and there is some evidence that similar mechanisms are responsible for the pain relief obtainable with periventricular stimulation in humans. Thus, it has been shown that such stimulation may cause an increase of β-endorphin in ventricular cerebrospinal fluid, but there is no proof that the activation of endophinergic systems is a prerequisite for the alleviation of pain. Naloxone has been extensively used to demonstrate that pain control with periventricular stimulation is dependent upon opioid mechanisms. It is true that this substance may reverse the effect of stimulation but this is not a regular finding. Also the phenomenon of tolerance to stimulation has been referred to in support of the activation of endogenous opioid systems. It should be emphasized, however, that there are reports on occasional patients who have retained the pain relieving effect of stimulation for many years without showing signs of tolerance.

Because of the profound insensitivity to noxious stimuli observed in animals during stimulation in the upper part of the brain stem, it is not unexpected that activation of the corresponding region in humans is preferentially effective for nociceptive pain. The normal perception of acutely induced painful stimuli is preserved, but chronic pain may be effectively alleviated. This contradiction is difficult to explain. It suggests that experimental pain in animals and chronic pain in humans are subserved by partially different neuronal mechanisms. Not un-

til recently has this problem been recognized as being of fundamental importance for the improvement of our methods of
dealing with chronic pain.

Further reading

Meyerson BA (1983): Electrostimulation procedures: Effects, pre-
sumed rationale, and possible mechanisms. In: *Advances in Pain
Research and Therapy* Vol 5, pp 495–534, Bonica JJ, et al, eds.
New York: Raven Press

Young RF, Feldman YA, Kroening R, Wayne F, Morris J (1984):
Electrical stimulation of the brain in the treatment of chronic pain
in man. In: *Advances in Pain Research and Therapy*, Vol 6, pp
289–303, Kruger L, Liebeskind JC, eds. New York: Raven Press

Pain, Neurophysiological Mechanisms of

Patrick D. Wall

Pain was classically defined as the sensation evoked by injury. This attractively simple definition neglects the observed phenomena of behavior and physiology and was replaced by the International Association for the Study of Pain in 1979 with the following definition and with a crucial coda:

> Pain is an unpleasant sensory and emotional experience associated with actual or potential tissue damage, or described in terms of such damage. Pain is always subjective. Each individual learns the application of the word through experience related to injury in early life. It is unquestionably a sensation in a part of the body but it is also always unpleasant and therefore also an emotional experience.

Many people report pain in the absence of tissue damage or any likely pathophysiological cause, usually this happens for psychological reasons. There is no way to distinguish their experience from that due to tissue damage, if we take the subjective report. If they regard their experience as pain and if they report it in the same ways as pain caused by tissue damage, it should be accepted as pain. This definition avoids tying pain to the stimulus. Activity induced in the nociceptor and nociceptive pathways by a noxious stimulus is not pain, which is always a psychological state, even though we may well appreciate that pain most often has a proximate physical cause.

This new definition, based on the experience and thoughts of many clinicians and scientists, sets up a severe challenge to the neurophysiologist who wishes to discover a mechanism related to pain. It is no longer sufficient to detect cells responding to noxious stimuli. Before launching in to recording or other experiments, it is first necessary to define the phenomena to be investigated.

The psychophysics of the relation of injury to pain

Both in the laboratory with controlled, defined stimuli and in the clinic, a highly variable relationship is observed between the stimulus and the response. Pain differs radically from such sensory experiences as the detection of light or sound where the great majority of the population obey predictable rules. This does not mean that the relation of injury to pain is mystical or random. It does mean that additional factors must be considered as well as the injury. These factors are now being defined. It warns the physiologist to expect the insertion of variable controls along the route from the injury to the sensory experience of pain. Furthermore a study of the actual phenomena shows that the variability is not restricted to the verbal statements of the patient or the general behavior of an animal. Many of the variations are equally apparent in autonomic responses such as blood pressure and in local segmental reflexes such as the flexion reflex. This fact was one of the general reasons to investigate the spinal cord for the possibility that

controls might exist at the entry point, the gate, where afferent impulses encountered the first central synapse as well as to investigate further stages of the transmission system.

The psychophysical phenomena after injury fall into at least three time epochs that alert the neurophysiologist to seek sequentially changing mechanisms. Attention is naturally directed to the acute phases within seconds after injury, which have been the main target for laboratory tests using brief, near noxious stimuli. Psychological and autonomic responses depend on the stimulus in context. The context includes other events in progress in the body, the brain set, expectation, previous experience, etc. Some 40% of patients admitted to a hospital emergency service with serious acute injury report no pain at the time of injury. The characteristics of this emergency analgesia are as follows: (1) It is instantaneous; there is no initial pain which is later brought under control. (2) It is localized precisely to the point of injury; subsequent trauma such as inserting a venepuncture needle evokes pain. (3) It is temporary; by 12 hours most patients experience the expected degree of pain. (4) An apparently similar phenomenon is observed in injured animals. Some 40% of normal people respond with a satisfactory analgesia to an inert placebo, and this analgesia has the following characteristics: (1) expectation or conditioning is necessary; (2) autonomic and somatic reflexes are affected as much as the verbal response; (3) repeated placebos habituate unless reinforcement is applied; (4) similar effects are observed in experienced animals.

The second phase of pain response may succeed tissue damage within minutes. The common experience of a twisted ankle exemplifies this where the instant acute pain dies down and is followed by a spreading, deep, poorly localized pain with tenderness of a quite different quality. Some of this second phase is explained by local changes that alter the nerve signals from the area of damage and are part of the local inflammatory reaction of redness and swelling. In addition, the effect is widespread in area and involves the motor system and changes in spinal cord circuitry.

The third phase has a latent period of days and is particularly apparent where nerve has been damaged or is diseased as in amputations or nerve diseases such as herpes zoster or diabetes. Here also there may be a peripheral component where the damaged nerves change their properties, but there is also a central reactive component. An important characteristic of the second and third phases is the appearance of tenderness (allodynia) where an innocuous stimulus applied to normal tissue at a long distance from the injury induces pain so that there is a complete uncoupling of the expected relationship of injury and pain.

Neurophysiology of peripheral nerve fibers

Single-unit studies in both humans and animals show that specific nociceptors exist among the small myelinated or A delta fibers and among the unmyelinated or C fibers. These fibers respond to relatively intense pressure, temperature, or chemical stimuli. The commonest type of C fiber, the polymodal nociceptor, responds to all three stimuli. The development of microneuronography in humans allows a relation to be established between the firing of specific types of fiber and the sensation experienced by the subject. The results show unequivocally that the experienced sensation is determined by central processes that take into account the context of the stimulus in addition to the evoked afferent barrage. The properties of sensation include threshold, intensity, quality (nature or modality), location, time course, and area. None of these sensory properties is the property of the information inherent in the discharge of a single type of afferent fiber. The sensory consequences are evidently the result of central analysis that takes into account the afferent barrage evoked by the noxious stimulus, plus the state of other aspects of the afferent barrage, plus the state of central synapses determined by central processes. Peripheral fibers are specific in relation to the stimulus that evokes activity in them, but they are not specific in relation to the sensory consequences of the presence of nerve impulses in a stimulus-specific fiber. Even the specific stimulus-response relationship of a single peripheral fiber changes in time after the first noxious stimulus. Some increase and others decrease their sensitivity.

Neurophysiology of spinal cord cells

Classical theory requires only the identification of cells that respond to activity in nociceptive afferents. Such cells have been detected in two areas of the dorsal horn that appear to be in monosynaptic contact with the nociceptive afferents. Necessarily there are many more cells excited indirectly. Direct contact is established in the most dorsal part of the dorsal horn in the substantia gelatinosa (lamina I and II) and also in the middle of the dorsal horn (lamina V). However, all cells thus far detected have two crucial properties in addition to their excitation by nociceptors: (1) they are inhibited by low threshold afferents, and (2) they are controlled by descending systems from the brain stem. The first has theoretical and practical consequences. For example, it means that the threshold of afferent firing produced by pressure needed to evoke pain is far higher than the threshold for temperature because an intense pressure stimulus necessarily fires low-threshold as well as high-threshold afferents while a temperature stimulus does not evoke activity in low-threshold mechanoreceptors. This effect was one of the reasons for my introduction of the now common pain therapy of transcutaneous electrical nerve stimulation, TENS. The descending systems controlling the response of cells to a nociceptor input originate particularly from the medullary and pontine reticular formation and from the raphe nuclei, which in turn are fired by other parts of the brain, particularly the periaqueductal gray matter. They provide a pathway by which the brain set can express itself by controlling the input transition from the afferents to the first central cells, the gate.

Although a small minority of dorsal horn cells are excited only by nociceptor afferents, the great majority are also excited by low-threshold afferents (wide dynamic range cells). Classical theory would predict that sensory experience of pain would result from the firing of the nociceptor-specific cells, but the experimental evidence suggests thus far that it relates best to cells that also signal innocuous events.

A sequence of pain mechanisms rather than a single one provides the basis for acute pain which evolves into chronic states. It is now becoming apparent that, after the initial arrival of an abnormal afferent volley that passes through a rapidly acting gate control, a subsequent series of changes alter the peripheral and central detection mechanisms. The delivery of an afferent volley in unmyelinated sensory fibers triggers both immediate and long-latency (several minutes) changes in the excitability of central synapses. These long-latency changes have the effect of opening central synapses that were not normally effective and in expanding receptive fields and in lowering thresholds. It is suspected that these long-latency, long-duration changes might provide a function for peptides characteristic of C afferent fibers, but proof of this speculation is not yet available. It is clear that these prolonged changes are particularly effectively triggered by impulses originating from deep tissue rather than from skin.

When nerves are damaged in the course of injury or disease, much longer latency changes in pain mechanisms result with delays of days and take weeks and months to reach their peak. Some can be attributed to changes in nerve membrane at the site of damage. Others occur centrally and produce changes of connectivity between afferents and central cells. They involve particularly the removal of normal inhibitions and may explain the excessive reactions seen in deafferentation syndromes. These central changes are not produced by nerve impulses but by the transport of chemical substances in afferent nerve fibers. C fibers appear to play a particularly important role in controlling central connectivity by way of transport. It remains to be discovered if the central changes are induced by a lack of normally transported substances such as nerve growth factor (NGF) or by delivery of abnormal compounds picked up from the damaged or diseased tissue.

Pain is not uniquely the result of the presence of activity in a stimulus-specific group of nociceptors. Even with acute, sudden, brief, unexpected injury, the sensory consequences depend on gate control affected (1) by the induced afferent barrage, (2) by other peripheral stimuli, and (3) by the activity of central structures. Injury induces prolonged changes in the periphery and in the spinal cord by way of nerve impulses that trigger connectivity changes with a latency of minutes and with a prolonged duration. Later changes by way of transport mechanisms induce further changes of central connectivity that again change the nature of pain mechanisms.

Further reading

McMahon SB, Wall PD (1985): Microneuronography and its relation to perceived sensation: A critical review. *Pain* 21:209–229

Melzack R, Wall PD (1982): *The Challenge of Pain*. New York: Basic Books

Wall PD (1984): Introduction. In: *The Textbook of Pain*, pp 1–16, Wall PD, Melzack R, eds. Edinburgh: Churchill Livingstone

Pain, Phantom Limb

William H. Sweet

Phantom limb pain refers to pain in two situations. In the first, a limb has been amputated accidentally or surgically, and the person continues to feel as though some or all the nonexistent limb is still present and hurting. In the second, most frequent when the living limb is largely denervated, the person feels as though the limb is in a different position from that of the actual structure. This occurs with denervation by injury to enough peripheral nerves, with avulsion of the nerve rootlets from the cord as in major brachial plexus injuries, with total transverse lesions of the entire cord in paraplegics, and finally, but much less often, with thalamic lesions. The manifestations in the two situations are different, and this discussion is confined to the first group who have had actual amputation. Ambroise Paré described "a thing wondrous strange and prodigious . . . patients who have many months after the cutting away of the leg, grievously complained that they yet felt exceeding great pain of that leg." Although nearly 100% of amputees have sensations referred to the missing limb, there is gross disagreement on the percentage who have frank pain. Three recent studies give 78% of 2694 patients, 72% of 58 patients, and 67% of 73 patients as having such pain. The huge series was drawn at random from the 25,000 U.S. veterans whose amputations were "military related." The results "make it clear that phantom pain severe enough to cause considerable discomfort and disruption of normal life is the usual condition for amputees." Nonpainful phantom sensations appear at once; there is usually a delay before these become painful. Although the pain subsides in some patients, once developed it usually persists. While often described as knifelike, sticking, or jabbing, many other adjectives have been used. Both constant and episodic varieties occur. Pain may be referred only to the stump, only to the phantom, or to both areas. The reason for these differences is unknown.

The longstanding debate continues regarding the mechanisms responsible for the pain. These appear to include not only peripheral neuropathy but also myelopathy and possibly as well reverberating abnormal circuits involving the brain. That input from the cut nerves in the stump plays a major role is proved by elimination of the pain in most patients given local anesthesia. In a small minority this relief outlasts the temporary objective sensory loss. The converse is also true, i.e., many types of local stimulation to the stump aggravate both painless and painful sensations in the phantom. Wall and Gutnick have demonstrated in rats that the neuroma at the proximal end of a cut sciatic nerve gradually generates a constant electrical potential demonstrable by recording from the appropriate posterior rootlets and blocked by local anesthesia to the neuroma. Such potentials may be responsible for pain and other sensations referred to the stump or the phantom. Persistent recurrence of these neuromas after multiple proximal neurectomies may be why such operations fail. Extensive posterior rhizotomies have also usually failed to give relief, possibly because of the conduction of some pain via ventral roots, demonstrated extensively in cats and in a few human patients as well.

Operations on the cord have sought to destroy pain activator mechanisms: (1) in the dorsal part of the dorsal horn of the cord for the full length of the zone related to the lesion or (2) in the contralateral ascending pain and temperature pathway for the secondary afferent neuron—the time-honored antero-lateral cordotomy.

The former operation—the dorsal root entry zone or DREZ lesion pioneered by Nashold—has proved especially valuable for pain that follows avulsion of the roots of the brachial plexus right out of the cord. Pain in the paralyzed upper limb, and persisting in the phantom that follows upon a futile amputation, is likely to be relieved. In this special type of trauma the lesion is clearly in the dorsal part of the cord.

Cordotomy usually stops amputation stump and phantom pain with a recurrence rate months or years later of about 50% in the originally successful cases. This is correlated with a regeneration of the capacity to feel a pin as sharp and does not argue in favor of a more rostral mechanism for the pain. However, the failure of these procedures when full analgesia is produced argues in favor of other pain pathways and mechanisms elsewhere in the cord or brain.

Further reading

Carlen PL, Wall PD, Nadvorna H, Steinbach T (1978): Phantom limbs and related phenomena in recent traumatic amputations. *Neurology* 28:211–217

Jensen TS, Krebs B, Nielsen J, Rasmussen P (1983): Phantom limb, phantom pain and stump pain in amputees during the first 6 months following limb amputation. *Pain* 17:243–256

Sherman RA, Sherman CJ, Parker L (1984): Chronic phantom and stump pain among American Veterans: Results of a survey. *Pain* 18:83–95

White JC, Sweet WH (1969): *Pain and the Neurosurgeon: A Forty Year Experience*. Springfield: Charles C Thomas, pp 68–86

Pain, Surgical Management of

Ronald R. Tasker

Introduction

Surgical treatment of pain should be considered only after all measures aimed at the direct causes of the pain have been exhausted, and the possible psychogenic aspects (hysteria, muscle tension, and magnification) fully evaluated. Surgical measures must be justified by the degree of interference with vocation, avocation, or sleep. Since no treatment is 100% effective and all threaten complications, the simplest measures should be considered first.

Nociceptive and deafferentation pain

Nonpsychogenic, intractable pain falls into two major categories: nociceptive and deafferentation syndromes. The former consists of chronic activation of nociceptors and the spinothalamic pathway, such as the pain caused by cancerous expansion of a long bone or compression of lumbosacral plexus. Patients referred to neurosurgeons for nociceptive pain usually suffer from cancer. Nociceptive pain is controlled by isolating this activity from consciousness. Deafferentation pain is usually dysesthetic or causalgic discomfort, sometimes associated with hyperpathia or sympathetic dystrophy, located in or adjacent to that part of the body whose subservient somatosensory system has been damaged to any degree at any level of the nervous system. Onset may be delayed after the causative injury, and the pain may be relieved by intravenous thiopental sodium but usually not by opiates. Though often relieved, like nociceptive pain, by proximal local anesthetic blockade, deafferentation pain is usually not relieved by surgical deafferentation at the same site. Though the pathophysiology is unknown, it is thought to be the result of some central process such as denervation neuronal hypersensitivity whether in instances of peripheral nerve damage, phantom pain, the pain felt below the level of spinal cord transection, or that caused by stroke. Patients referred to neurosurgeons for this kind of pain usually suffer from noncancerous conditions, though malignancy commonly produces deafferentation pain as well, perhaps accounting for the confusion in describing results of pain treatment.

Surgical treatment of nociceptive pain

Both destructive and modulatory techniques have been advocated for the treatment of nociceptive pain.

Destructive techniques. Surgical treatment of pain has increasingly used percutaneous techniques, particularly since the introduction of image intensification and radiofrequency lesion making (Fig. 1). The simplest destructive measures consist of percutaneous radiofrequency neurectomy or dorsal rhizotomy, particularly useful in the trigeminal, intercostal, and facet nerves of the spine because of the lack of disability from motor and sensory loss. But these convenient, precise, low-risk techniques are seldom indicated because of the propensity of nociceptive pain not to respect the provinces of individual nerves or roots. Usually more complex procedures are required. Percutaneous cordotomy by the lateral high cervical technique is the treatment of choice for nociceptive pain below the C4 dermatome with the following exceptions: (1) midline symmetrical pain caused by cancer is probably better treated by epidural or intrathecal opiate infusion; and (2) cord lesions producing high cervical levels of analgesia must be avoided ipsilateral to a solitary normally functioning lung either by virture of pulmonary, corticospinal, reticulospinal, or phrenic damage. Open cordotomy is seldom if ever necessary. Percutaneous cordotomy by the low anterior cervical technique may be useful if pathology interferes at higher levels or if bilateral cordotomy is necessary and a high cervical level of analgesia was achieved on the first side. For the 5% of patients (Fig. 2) requiring surgical treatment of nociceptive pain above the C5 dermatome and the occasional patient with other contraindications to cordotomy, many procedures are available but none is as satisfactory as cordotomy. Percutaneous upper cervical trigeminal tractotomy is less satisfactory than trigeminal rhizotomy. Alcohol injection of the pituitary is a simple inexpensive way to treat cancer pain but has a relatively low success rate (40–60%) and a high incidence of neurological complications even if they are mainly transient. Chronic stimulation of periventricular or periaqueductal gray (PVG-PAG) is a possibility. Stereotactic interruption of the spinothalamic tract at the high mesencephalic level is technically complex, carries a 9% major

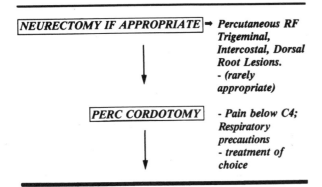

Figure 1. Flow sheet of surgical therapy for nociceptive pain below the neck.

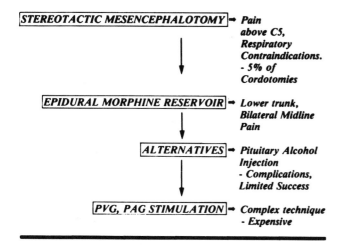

Figure 2. Flow sheet for nociceptive pain above neck and that not otherwise amenable to cordotomy.

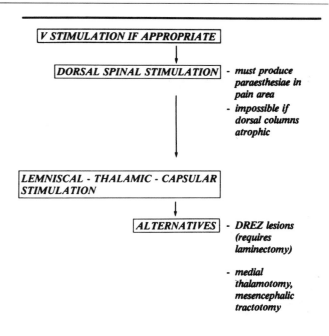

Figure 3. Flow sheet for surgical treatment of deafferentation pain.

and 30% minor risk but a relatively high (70–80%) chance of success. Stereotactic interruption of the (presumed) reticulothalamic system at either the high mesencephalic (higher morbidity) or the medial thalamic (lower success rate) levels yields a 60–70% success rate. Little information is available about pulvinarotomy, and hypothalamotomy and psychosurgical procedures, particularly cingulotomy, are seldom used today.

Modulatory techniques. For nociceptive pain only two modulatory techniques are in regular current use: chronic PVG-PAG stimulation and epidural or intrathecal opiate infusion. The former is technically complicated and requires expensive equipment but carries a less than 5% morbidity. It is otherwise well worth considering in patients not suitable for cordotomy. Opiate infusion requires equipment that is sometimes expensive, is prone to failure, and requires continual refilling. It is the preferred treatment for midline cancer pain. Chronic stimulation of peripheral nerves, cord, or other brain sites is not usually useful in nociceptive syndromes.

Surgical treatment of deafferentation pain (Fig. 3)

Destructive techniques. In general destructive techniques offer no more than a 25–30% chance of relief of deafferentation syndromes. After all, they only add to the deafferentation. Accompanying sympathetic syndromes, however, may be relieved by sympathectomy and hyperpathia (depending on incomplete denervation) by completing the deafferentation.

Modulatory techniques. Currently the treatment of choice for deafferentation syndromes is chronic stimulation. Of the peripheral nerve techniques, only trigeminal stimulation is attractive since its successful use obviates stereotactic brain stimulation. Dorsal column stimulation (DCS) is the preferred treatment for deafferentation syndromes. It is simple to execute percutaneously, carries a less than 1% risk of complications, and has the advantage of being easily assessed by transcutaneous trial prior to internalization of the system. Some 50–70% of candidates so tested report pain amelioration and can be

considered for chronic implantation. But the simpler percutaneously introduced electrodes are prone to migrate (25%), and the equipment is expensive and must be continuously serviced and manipulated by the patient. There is also a small annual fall-off in pain relief of a few percent per year despite continuing technical success. Though the mechanism by which chronic stimulation relieves deafferentation pain is unknown, it probably inhibits the abnormal central neural activity produced by the original deafferentation that is presumably responsible for the pain. For its success it must produce paresthesiae in the area of the patient's pain. In those cases in which DCS fails, it is often because appropriate paresthesiae cannot be produced when the dorsal columns have atrophied after lesions such as cord transection. Stereotactic chronic stimulation of the medial lemniscus-ventrocaudal nucleus-somatosensory capsule is the main alternative. The criteria are similar to those for DCS. For stimulation failures the DREZ operation is available though it requires a laminectomy and carries a significant morbidity.

Further reading

Miles J (1979): Chemical hypophysectomy. In: *Advances in Pain Research and Therapy, Vol 2,* Bonica JJ, Ventafridda V (eds). New York: Raven Press, pp 373–380

Hosobuchi Y (1980): The current status of analgesic brain stimulation. *Acta Neurochir (Suppl)* 30:219–227

Shetter AG, Hadley MN, Wilkinson E (1986): Administration of intraspinal morphine sulfate for the treatment of intractable cancer pain. *Neurosurgery* 18:740–747

Tasker RR (1982): Percutaneous cordotomy—the lateral high cervical technique. In: *Operative Neurosurgical Techniques: Indications, Methods, and Results.* Schmidek HH, Sweet WH (eds.). New York: Grune and Stratton, pp 1137–1153

Pain Management, Focal Instillation of Opiates in the Central Nervous System

Charles E. Poletti

The use of implantable systems for the long-term focal delivery of pharmacologic agents to the central nervous system (CNS) represents a major therapeutic advance. These systems can now be used to deliver opiates, hormones, neurotransmitters, antibiotics, anticoagulants, oncolytic agents, and other compounds targeted to the epidural space, cerebrospinal fluid, neuraxis, or bloodstream. In the management of pain states this approach has the advantage of avoiding both the systemic effects associated with high-dose narcotics and the problems associated with the surgical destruction of portions of the nervous system.

Spinal instillation of morphine

Opiate receptors, which are activated endogenously by met-enkephalin, have been discovered autoradiographically in spinal cord laminae 1 and 2 (substantia gelatinosa). Morphine has also been demonstrated to activate these receptors and has been shown to selectively inhibit the release of substance P, suppressing peripheral nociceptive input to the substantia gelantinosa. Spinal segmental analgesia, which is reversible by naloxone and unaccompanied by effects on motor or sensory function, has been elicited in the rat, cat, and primate with microgram doses of morphine in the spinal subarachnoid space. Based on this rationale, spinal morphine is now being used clinically both for acute pain and for chronic pain.

Acute postoperative pain is being treated, with increasing frequency, by focally administered morphine to the spinal cord via an epidural catheter. Small doses of morphine given after major leg, abdominal, or low thoracic operations often provide satisfactory relief from pain without making the patient sleepy or obtunded.

Chronic pain caused by cancer of the leg, abdomen, or lower thoracic region is now, in many centers throughout the world, also being treated by the direct application of morphine to the spinal cord—either from the epidural space or by instillation in the subarachnoid spinal cerebrospinal fluid. Patients who become obtunded by systemic opiates before obtaining pain relief can obtain satisfactory pain relief from 2–3 mg epidural morphine or 0.5–1 mg morphine instilled in the spinal subarachnoid space. These relatively low doses have minimal systemic effects, relieving the patients of pain while leaving them fully alert. While very effective in most cancer patients, spinal morphine does not relieve pain in all patients with cancer. Furthermore, because of the possibility of respiratory suppression, especially in debilitated patients, most centers are not applying morphine to the upper thoracic or cervical spinal cord in patients with cancer in these regions.

A few centers are using spinal opiates to treat chronic pain caused by benign diseases. The use of long-term spinal opiates for benign pain is usually limited to selected patients with severe lumbar arachnoiditis, phantom limb pain, postherpetic pain, and severe chronic pancreatitis.

Intracerebral instillation of morphine

Opiate receptors have also been found to be especially concentrated in the periaqueductal gray area, adjacent to the aqueduct running between the 3rd and 4th ventricles and the amygdala, adjacent to the ventricle in the temporal horn. Both structures have been implicated in the modulation of central nociception and even nociception at the spinal level via the descending analgesic system. Accordingly, the idea has developed of activating these cerebral opiate receptors by instilling morphine into the cerebral ventricular system. Such clinical trials are now being run in a number of centers throughout the world. The five centers with publications—totaling 96 patients—all report promising results.

Intraventricular morphine therapy seems especially appropriate in patients suffering from cancer of the head, mouth, throat, and neck, shoulders, arms, and upper thoracic regions. Over 90% of 96 reported patients had good or excellent pain relief, an even better result than that obtained with spinal morphine. Pain relief is obtained with only 0.1–2 mg morphine per day delivered via a lateral ventricular catheter.

Respiratory depression, clearing after the first few doses, has been seen in only 2 of the 96 patients, both markedly debilitated. Nausea, occasional vomiting, and anorexia occur in 25–50% of the patients; these occur during the first week or so of therapy and generally can be controlled with standard antiemetic medicines. Other side effects are minor and transient. Most notably, patients obtain pain relief without any obfuscation of their mental capacities.

Long-term, focal opiate therapy—either spinal or intracerebral—causes tachyphylaxis. Usually, this is not a problem in the cancer patients since the initial doses are low and longevity relatively short. This may, however, represent a serious limitation in the longer term treatment of benign intractable pain. In the future other agents will be developed to activate other opiate receptors and probably other pain-suppressing nonopiate receptors that do not have cross-tolerance with the subset of morphine-activated opiate receptors.

Two additional problems potentially plaguing the field of applied focal pharmacology—be it to the CNS or to other bodily systems—are (1) the instrumentation of the implanted delivery system and (2) the associated incidence of sepsis. Extensive work is under way to reduce these problems.

Further reading

Cousin MJ, Mather LE (1984): Intrathecal and epidural administration of opioids. *Anesthesiology* 61:276–310

Poletti CE, Schmidek HH, Sweet WH, Pilon RN (1982): Pain control with implantable systems for the long-term infusion of intraspinal opioids in man. In: *Operative Neurosurgical Techniques* 2:1199–1210

Yaksh TL (1981): Spinal opiate analgesia: characteristics and principle of action. *Pain* 11:293–346

Pain Measurement by Signal Detection Theory

W. Crawford Clark

Pain is a subjective experience with somatosensory, cognitive, emotional, and other components; yet, if we are to understand and treat pain, we must measure pain in a reliable and valid manner. To clarify these problems, various psychophysical models have been used to examine verbal reports to calibrated electrical, thermal, pressure, and other physical stimuli. These psychophysical procedures include the traditional threshold, Stevens magnitude estimation, various multidimensional scaling procedures, as well as signal detection, or more descriptively, sensory decision theory (SDT). The merit of the SDT approach resides in its unique ability to quantify separately sensory and attitudinal parameters. The index of discriminability, d' or $P(A)$, measures the accuracy (it is possible in the binary decision procedure to score the subject's response as right or wrong) with which an individual distinguishes among stimuli of various intensities. This index of sensory performance has been demonstrated to be essentially uninfluenced by changes in the subjects' expectation, mood, motivation, and other attitudinal variables. The discriminability index is related to the functioning of the neurosensory system: high values suggest that neurosensory functioning is normal, low values that the amount of information transmitted to higher centers is low. The other measure of perceptual performance is the report criterion (Lx or B), which measures response bias, that is, the willingness or reluctance of a subject to use a particular response. It is related to the subject's attitude toward the sensory experience: a high criterion reflects stoicism, while a low criterion indicates that the subject readily reports pain. Only SDT yields separate measures that can be so readily identified with the neurosensory and the emotional components of pain. Other psychophysical procedures mix these two components into a single measure: the threshold in the methods of limits and constant stimuli, and the slope or exponent in magnitude estimation procedures.

Introduction to the SDT model

An example of the stimulus-response matrix of sensory decision theory appears in Table 1. In the simple binary decision case, high-intensity (stimulus) and low-intensity stimuli (blanks), or more generally, any objectively definable events A and B, are presented randomly to the observer. The theory assumes that background interference, or noise, is always present in amounts that vary randomly over time. Thus, there are two overlapping distributions of information. The task of the observer is to decide whether the information sampled was produced by the higher or the lower intensity stimulus (or by event A or event B). The observer makes a statistical decision and responds "high" or "low," "painful" or "not painful," or "A" or "B." The four possible decision outcomes appear in Table 1. A plot of hit rate against false alarm rate yields the receiver-operating-characteristic (ROC) curve (see Fig. 1). Although, as is often done, more than two intensities of stimulation are presented, they are analyzed in pairs. The binary choice is usually extended to a rating scale of 8 to 16 categories, the number of points on the ROC curve being one less than the number of categories. The analysis of the data, however, is essentially the same as for the binary task.

Parametric and nonparametric models of SDT exist. The parametric SDT procedure assumes that the underlying noise and signal-plus-noise distributions are unit normal probability density functions and, for d', of equal variance. The distance between the means of these distributions in terms of Z, the standard deviate, indexes discrimination, d' (one value for each point on the ROC curve). If Z_N and Z_{SN} are the standard deviation measures of the distance of the criterion from the means of the noise and signal-plus-noise distributions, then,

$$d' = Z_N - Z_{SN}$$

Thus, to determine d' from the empirically established false alarm and hit probabilities, one proceeds to a published table that relates areas under the normal curve to values of the standard deviate, Z. The criterion (one for each point on the ROC curve) is a value of the logarithm of the likelihood ratio, Lx. Lx, the likelihood ratio criterion, is the ratio of the ordinate of the signal-plus-noise distribution to the ordinate of the noise distribution, at the criterion locus defined by hit and false alarm probabilities, respectively,

$$Lx = f_{SN}(y)/f_N(y)$$

Accordingly, to determine Lx from the false alarm and hit probabilities, one simply consults a table that relates areas under the normal curve to values of the ordinate.

Table 1. Stimulus-Response Matrix for Binary Decision Task

		Response	
		"Pain," or "High," or "A occurred"	"No Pain," or "Low," or "B occurred"
Stimuli	Higher Intensity Stimulus or Event A	Hit	Miss
	Lower Intensity Stimulus or Event B	False Alarm	Correct Rejection

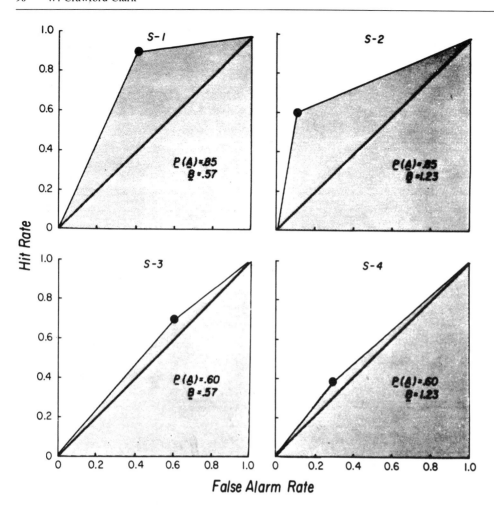

Figure 1. Receiver-operating-characteristic (ROC) curves for four subjects with high (S-1, S-2) and low (S-3, S-4) discriminability, and high (S-2, S-4) and low (S-1, S-3) report criteria.

McNicol has presented an excellent introduction to the nonparametric SDT model. Here, $P(A)$, the area under the ROC curve (see Fig. 1) indexes discrimination. To compute $P(A)$ the point(s) on the ROC curve including (0,0) and (1,1) are joined by straight lines and the areas of the triangle and trapezia are summed. Because $P(A)$ integrates all points on the ROC curve, it is less variable than d', particularly when the number of observations are small. The criterion, B, is defined as the rating scale category at which the cumulated hit-plus-false-alarm probabilities equal unity. Equivalently, B locates the median, that is, the rating scale criterion at which half the responses (to both stimulus intensities) are to higher response categories and half are to lower. Both the parametric and nonparametric procedures are explained in a simple, scholarly way by McNicol. In pain research, where the number of presentations of stimuli must, perforce, be relatively low, the discrimination index of the nonparametric SDT model is preferable to the parametric. The criterion B is more stable than Lx; however, it may miss interesting shifts in the criterion at the extreme categories. It is safer to compute both. Because $P(A)$ and Lx are not normally distributed, they should be subjected, respectively, to 2 arcsin square root $P(A)$ and logarithmic transformations for statistical analysis.

Unlike the other procedures, a third SDT model and procedure recently developed by Buchsbaum and co-workers yields a measure of discriminability based on information integrated over any number of stimulus intensities, as well as any number of report categories. The wider range of stimulus intensities that can be incorporated into a single discrimination index gives this procedure a flexibility which should prove useful in clinical studies, where pain sensitivity often varies widely among patients, and time allows but a small number of observations.

The discriminability and response bias parameters of SDT may be understood by viewing the stimulus-response matrices of subjects who differ in their ability to discriminate between higher and lower intensity stimuli, and who differ in where they locate their subjective criterion for reporting pain (see Fig. 1). Subjects S-1 and S-2, who have relatively high hit rates in combination with low false alarm rates, are superior discriminators, high $P(A)$, compared to S-3 and S-4. Note that $P(A) = 0.5$ is chance performance, or zero discrimination. Subjects S-1 and S-3, for whom the sum of the hit rates and false alarm rates is high (many pain reports), have set a low pain report criterion, while S-2 and S-4 have set a high, or stoical, pain report criterion. It is important to note that the hit rate alone, which is equivalent to the threshold of traditional psychophysics, gives quite a different (and erroneous) view of pain sensitivity. S-1 (hit rate = 0.9) appears to be most sensitive to pain, followed by S-3 (0.7), S-2 (0.6), and S-4 (0.4). But appearances are deceptive; for SDT demonstrated that S-1 and S-2 were equally good discriminators, while S-3 and S-4 were equally poor discriminators. This example demonstrates that the threshold of classical psychophysics is an unanalyzable amalgam of sensory sensitivity and psychological or attitudinal report bias, and cannot be trusted as an index of sensory functioning.

Empirical findings

Criterion changes. A placebo described and accepted as a powerful analgesic markedly raises the pain threshold, that is, apparently decreases the sensitivity to noxious thermal stimulation. However, SDT has demonstrated that this decrease in the report of pain is caused by a raised pain report criterion and not by a decrease in thermal discriminability. Since analgesics such as morphine have been shown to decrease discriminability, but the placebo was without effect, it may be concluded that the placebo-induced reduction in pain report (raised threshold) was not due to analgesia, but was due to a criterion shift made in response to the social demand characteristics of the experimental situation. Other studies have shown that directed suggestion can either raise or lower the pain report criterion; that older individuals, certain ethnocultural groups, and males (compared to females) set higher pain report criteria; and that anxious subjects and those manifesting certain personality characteristics set low pain report criteria. In each of these studies the various experimental manipulations or psychological differences were without effect on the discrimination parameter. Thus, these differences in pain report did not reflect differences in underlying neurosensory activity related to the sensation of pain, but only differences in the subjects' criterion for reporting pain.

Discrimination changes. Analgesics would be expected to decrease discriminability by attenuating the amount of neurosensory information reaching higher centers. In addition, the pain report criterion might also be raised, since if less pain is felt, less pain should be reported. These effects on discriminability and criterion have been found with a number of analgesic agents including nitrous oxide, morphine, and codeine. Although it would be wrong to equate decreased discriminability with analgesia in every instance, there is mounting evidence that the SDT discriminability index is a frequent correlate of analgesia and may prove useful in drug research. For example, certain putative analgesics possess mood-altering properties; thus, it is possible that improved mood (happy patients report less pain), rather than an analgesic effect per se, is responsible for some of the analgesic effects reported clinically. For example, cannabinoids have been thought to produce analgesia, however, the evidence from an SDT study suggests that patient mood but not pain was altered. Other studies have demonstrated that SDT parameters are sensitive to changes in endocrine and brain enkephalin levels. Both chronic back pain patients and depressed patients have been shown to have poor discriminability and to set a stoical pain report criterion.

Other applications. SDT is not simply a model for extracting signal from noise, as those who are but superficially acquainted with the model contend. Actually, from its inception SDT has been a general decision-making model: either event A or event B may occur, and the observer, on the basis of partial information, must decide which event has occurred. These events might be simple physical stimuli or more cognitive, such as pathological and normal X-rays, the likelihood of a symptom being present in disease A and in disease B, success and failure rates with treatments A and B for an illness, or the side effects and target effects of a drug. Nor need the responses be verbal: autonomic responses, evoked potentials, and avoidance responses in animals have been studied with the SDT paradigm.

Conclusions

The failure to quantify pain has seriously handicapped our ability to relieve it; without measurement, it is impossible to compare accurately the efficacy of the various treatments available for the relief of pain. One reason for this dismal record may be the mistaken belief that the report of pain is a precise description of a sensory experience, uncolored by expectations, attitudes, emotions, etc. Among the psychophysical models now available, only SDT permits the independent quantification of the sensory and attitudinal experiences.

Further reading

Buchsbaum MS, Davis GC, Coppola R, Naber D (1981): Opiate pharmacology and individual differences. I Psychophysical pain measurements. *Pain* 10:357–366

Clark WC (1974): Pain sensitivity and the report of pain: An introduction to sensory decision theory. *Anesthesiology*, 40:272–287. Reprinted in Weisenberg M, Tursky B, eds (1976): *Pain: New Perspectives in Therapy and Research.* New York: Plenum Press, 195–222

Clark WC, Yang JC (1983): Applications of sensory decision theory to problems in laboratory and clinical pain. In: *Pain Measurement and Assessment.* Melzack R, ed. New York: Raven Press 15–25

McNicol D (1972): *A Primer of Signal Detection Theory.* London: Allen & Unwin

Paraneurons

Tsuneo Fujita

What are paraneurons?

Paraneurons (a term originated by Fujita in 1975) are groups of cells which have not been classified as neurons and yet share certain morphological and functional features with neurons. Paraneurons are defined by three criteria: (1) They possess neurosecretion-like and/or synaptic-vesicle-like granules. (2) They produce substances identical with or related to neurosecretions or neurotransmitters. (3) They recognize adequate stimuli and respond to them by the release of their secretions. As item (3) indicates, paraneurons can be characterized as receptosecretory cells. If the receptive function of a cell appears dominant, it can be called a sensory paraneuron. If the secretory function comes to the front, the cell can be called a secretory or endocrine paraneuron.

No clear boundary can be drawn between neurons and paraneurons; both comprise a spectrum of cells (Fig. 1). Cytological features at the light and electron microscopic levels, as well as mechanisms of stimulus-secretion coupling are common to neurons and paraneurons. Some paraneurons show action-potential-like depolarization of the membrane when stimulated. Recent immunocytochemical studies show the occurrence of neuron-specific proteins (neuron-specific enolase; neurofilament protein; Purkinje cell specific protein, etc.) in many paraneurons. Some paraneurons are supported by cells immunoreactive for glia-specific protein, S-100. Some paraneurons are synaptically connected with neurons, forming neuroparaneuronal chains, whereas others are free from nervous supply.

Paraneurons are multiple in origin. They may be neuroectodermal, nonneural but nerve dependent, or nonneural and nerve independent. Convergence differentiation is believed to have produced the cell group of paraneurons from variable origins.

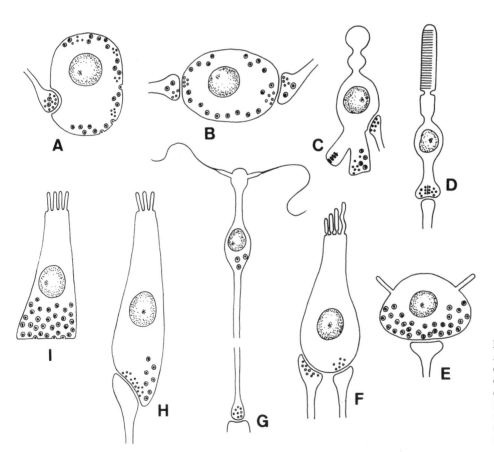

Figure 1. Main types of paraneurons. A. Adrenal chromaffin cell and some other endocrine cells. B. Carotid body chief cell and some other internuntial cells. C. Pinealocyte. D. visual cell. E. Merkel cell in the skin. F. Hair cell in the inner ear and lateral line organ. G. Olfactory cell. H. gustatory cell in the taste bud. I. Gastroenteric endocrine cell.

Table 1. Main Paraneurons and Their Peptidic and Aminic Products

Paraneurons	Peptides	Amines
Endocrine cells		
Adrenomedullary cell	enkephalins, NPY	Adrenalin, noradrenalin
Carotid body cell	enkephalins	dopamine
Thyroid parafollicular cell	calcitonin, somatostatin	serotonin, dopamine
Parathyroid cell	parathormone	?
Adenohypophyseal cell	GH, ACTH, TSH, LH, FSH, etc.	dopamine (in some cells)
GEP endocrine cell	insulin, glicentin/glucagon, gastrin, secretin, CCK, somatostatin, enkephalins, etc.	serotonin (in some cells), dopamine (in some cells)
Bronchial endocrine cell	GRP	serotonin
Pinealocyte	?	melatonin
Sensory cells		
Retinal visual cell	?	catecholamine ?
Inner ear hair cell	?	catecholamine ?
Merkel cell	enkephalins ?	?
Gustatory cell	?	serotonin
Olfactory cell	?	acetylcholine ?

Secretions of paraneurons

Bioactive peptides seem the most important elements in the contents of paraneuronal granules. Usually, a large prepropeptide precursor is synthesized and processed in Golgi complexes and maturing granules into smaller fragments, some of which are bioactive. As numerous bioactive peptides are shared by neurons in the brain and by paraneurons in the gut and pancreas, the term brain-gut peptides has become popular. The precursors of the peptides and the mechanisms of their processing are common to neurons and paraneurons in most cases.

Moreover, other categories of granule substances are largely common to neurons and paraneurons. The best known are bioactive monoamines such as noradrenaline, adrenaline, dopamine, and serotonin (5-HT). The APUD concept of A.G.E. Pearse is based on the occurrence of amines in both neurons and paraneurons. Amines, however, are detectable only in some of those cells.

More constant elements in the granules of neurons and paraneurons are adenosine triphosphate and other adenine nucleotides. Acidic glucoproteins, represented by chromogranin A (in adrenal chromaffin cells) are believed to serve as the carrier proteins of other smaller molecules. Chromaffin granules and cholinergic synaptic vesicles have been shown to contain highly negatively charged mucopolysaccharides resembling heparin, and other granules of neurons and paraneurons are also presumed to contain this kind of substance. Metachromatic reactions of neuroparaneuronal granules are ascribed to the occurrence of these glucoproteins and mucopolysaccharides. High concentrations of Ca^{2+} are also characteristic of neuroparaneuronal granules.

Small synaptic vesicles seem to be exceptions in the sense that they neither contain peptides nor glycoproteins. Besides

Bioactive peptides and amines identical with or closely related to those in vertebrates are found in coelenterate and higher invertebrates, suggesting that they are used as neuronal and paraneuronal messengers throughout the evolution of multicellular animals.

Neoplasm

Neoplasm of the paraneuron is called paraneuroma or, in certain cases, paraneuroblastoma. It includes Pearse's apudoma. There is every gradation between the paraneuroma and the neoplasm of the neuron (neuroma, neuroblastoma, etc.). Various neuronal and paraneuronal cells may be simultaneously involved in neoplastic or hyperplastic changes (multiple endocrine adenomatosis).

Further reading

Fujita T (1977): Concept of paraneurons. *Arch Histol Jpn* 40:1–12

Fujita T, et al. (1979): Current views on the paraneuron concept. *Trend Neurosci* 2:27–30

Fujita T (1983): Messenger substances of neurons and paraneurons: their chemical nature and the routes and ranges of their transport to targets. *Biomed Res* 4:239–256

Fujita T, et al. (1983): Immunohistochemical detection of nervous system-specific proteins in normal and neoplastic paraneurons in the gut and pancreas. In: *Gut Peptides and Ulcer*, Miyoshi A ed. Tokyo: Biomedical Research Foundation

Pheromones

Eric Barrington Keverne

Olfactory stimuli play an important part in animal communication, and a special terminology has been used for substances that function in this manner. In 1932 Bethe employed the term *ectohormone* with reference to substances, such as insect sexual attractants, which are external rather than internal secretions and act as chemical messengers between individuals rather than between different parts of the same individual. This self-contradictory term went out of use following the introduction of the term *pheromone,* defined by Karlson and Butenandt in 1959 as "substances secreted to the outside of an individual and received by a second individual of the same species in which they release a specific reaction, for example, a definite behaviour or developmental process."

Considerable discussion has surrounded the definition of the term pheromone. In the early 1960s Wilson considered two classes of chemical signal: releaser and primer pheromones. Releaser pheromones stimulate an immediate behavioral response and may be further classified as to their supposed function such as sex attractants, alarm pheromones, aggregation pheromones, or trail marking pheromones. Primer pheromones do not induce a behavioral response but result in a physiological change in the receiver. The queen bee substance is an example of a primer pheromone that inhibits ovarian development in worker bees, while male mouse urine contains primer pheromones that accelerate the onset of puberty in young females, induce estrus in adult females, and cause implantation failure in recently mated females.

For a large variety of insects, our present knowledge of the diversity, complexity, and specificity of the pheromones is well established. In the early days of pheromone chemistry, it was believed that every insect had its own specific and characteristic pheromone for a certain chemical message: one compound for attracting, the other for marking an odor trail, and yet another for aggregation. Now, however, we know that not only may one compound act as a pheromone in several different species, but the pheromone may only elicit a certain response in the presence of at least one other compound. Such synergisms have probably formed the basis of reproductive isolating mechanisms, which have been of importance for the evolution of a species. Synergism among components of an insect's secretion appears to be a general phenomenon and may take the form of synergism among isomeric components, homologs differing in the number of carbon atoms, or mixtures of several aliphatic acids, or alcohols. In such multicomponent pheromones, minor components can be extremely important, and chemical messages between insects are not simple, one-syllable words. Even the first pheromone to be chemically identified, the unsaturated alcohol bombykol, is now known to be but one component of a complex secretion requiring a geometric isomer of bombykol as well as the corresponding aldehyde for optimal effectiveness. The effectiveness of insect pheromones is quite remarkable, and their attractiveness over great distances has been put to commercial use for selective eradication of insect pests.

Receptors

Studies of chemoreception at the molecular level have shown that proteins function as receptor sites. When an odorant chemical reaches the receptor site, it is commonly supposed that a change in conformation of the receptor site or association with the protein receptor molecule alters the ionic permeability of the receptor cell membrane. This gives rise first to a generator potential that, when strong enough, is succeeded by an action potential or spike. The cilia of olfactory receptors are believed to contain the receptor sites for different odor molecules, and electron microscopy of freeze-fracture preparations reveals membrane particles that are appreciably more concentrated than on adjacent motile respiratory cilia.

The technical difficulties imposed on recordings from the olfactory receptors in the nose of mammals, and the relatively few identified mammalian pheromones, restricts our understanding of pheromone receptors to insects. Some receptors are designed for the detection of a single compound and may be highly selective, not even responding to chemical isomers of the pheromone unless present in concentrations a 1,000-fold greater than the natural pheromone. In contrast to these odor specialists, other receptors, odor generalists, may respond to more than one compound. They may depolarize and show an increase in impulse firing to some compounds or hyperpolarize and be inhibited from firing by other compounds. Some odor molecules may have more than one active site that interacts with different binding proteins. If these binding proteins vary in their affinity for the molecule and are distributed along the membrane of different receptor cells, then even a pure odor is capable of generating different patterns of receptor activity according to the concentration. This phenomenon may also account for the fact that some odors smell differently to humans at increasing concentrations. By introducing the idea of multiple recognition sites for a pheromone complex, we can imagine a coding system at the receptor level of immense specificity. Likewise, if a receptor cell requires two, three, or more sites to be bound at the receptor molecule (allosteric receptors) before it fires, this permits a totally different firing pattern for a pheromone complex over and above that produced by each of the component compounds. Only the correct proportions in the mixture provide the correct signal. A mixture of differing proportions may even be the signal for another species. Hence the complete information about a pheromone complex (and very few simple pheromones exist, even in insects) is coded in a complicated pattern of neural signals in many channels to the central nervous system rather than an all-or-none activation in one or a few discrete channels. Hence, if we wish to understand the action of pheromones, it is necessary to follow the coded receptor signal or impulse pattern synapse by synapse until we end up in higher integrative centers of the brain, or in the case of insects, that area of the central nervous system which integrates with motor output.

Neural basis of pheromone action in mammals

If we consider the way olfactory information is transmitted centrally to induce a behavioral event, then at least two separate neural pathways have to be taken into account. The olfactory pathway that conveys information that results in one animal being attracted to another, for example, is clearly of a diffuse nature, since attraction in this sense can be modified by a number of variables, including past experiences and partner preferences. Such olfactory information requires complex integration and is probably represented neuroanatomically by olfactory pathways that enter the pyriform cortex and are relayed to the thalamus and on to the prefrontal area of the neocortex, before finally gaining access to the anterior hypothalamus, that area of the brain primarily concerned with sexual behavior (Fig. 1). In taking this indirect route, access to the neocortical neuron pool is achieved, and olfactory information may be integrated and modified before being relayed to that area of the brain responsible for coordinating sexual behavior. Hence, such olfactory cues would only lead to the onset of sexual arousal on appropriate occasions, e.g., with an attractive, receptive female and in the absence of dominant males or predators. Not only does this allow for a degree of plasticity in the behavioral response to pheromones, but it also accounts for the kind of variability observed in response to sex attractants among many mammals. If odor cues were to release a fixed action pattern of behavior, as seen among insects, this could well constitute a threat to the species (although certain mammals may behave irrationally in the presence of attractive females, none find themselves transfixed in a sexual limbo and unable to escape approaching predators).

Most mammals, the exceptions being higher primates and humans, possess a well-developed dual olfactory system. The neural pathway which produces the kind of response described above is called the main olfactory pathway and has its receptors located in the nasal cavity. These give rise to axons, the fila olfactoria, which ascend into the cranium through the cribriform plate to synapse on mitral cells in the olfactory bulbs. A second set of olfactory receptors, located in the epithelial lining of the vomeronasal organ, also sends axons, the vomeronasal nerve, through the cribriform plate. Vomeronasal nerves run medially between the main olfactory bulbs and synapse in the accessory olfactory bulbs. The accessory olfactory bulb itself is spatially and histologically distinct from the main olfactory bulb and usually lies on the dorsal surface of the main bulbs. The vomeronasal pathway has given rise to the idea

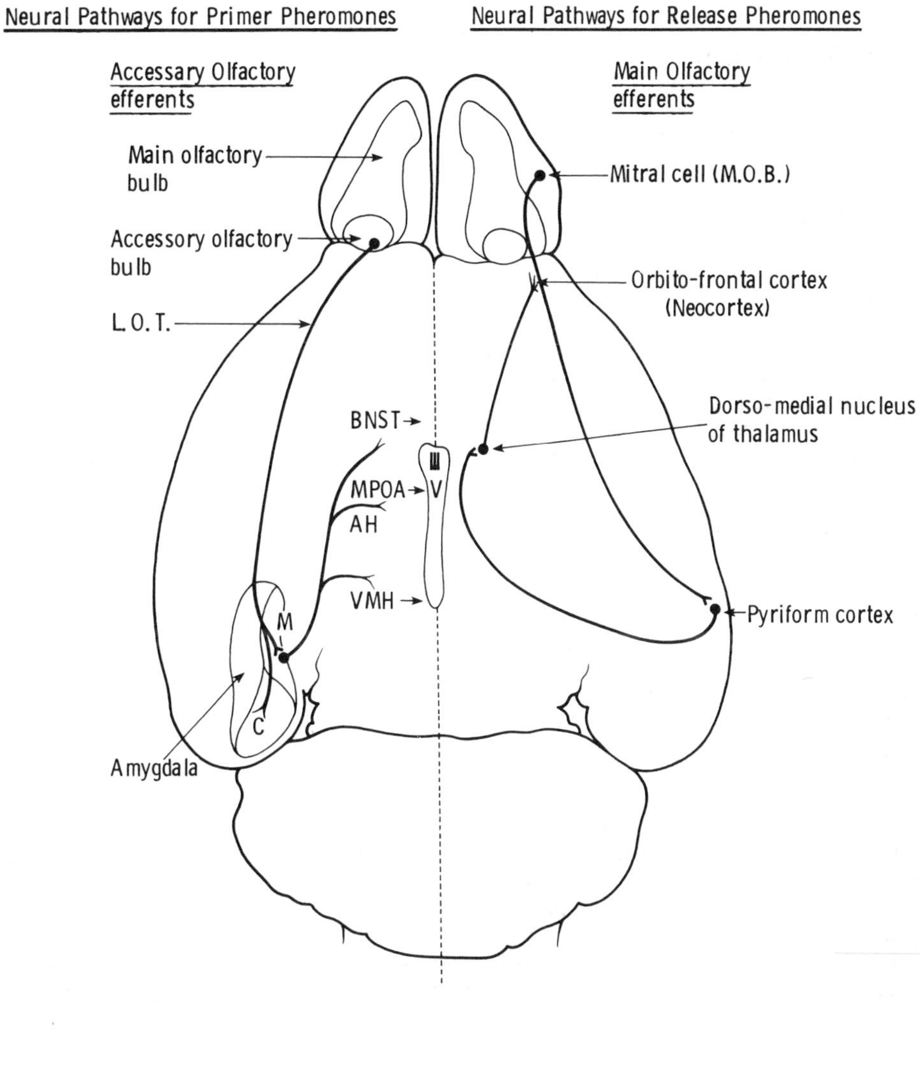

Figure 1. Pheromone pathways. See text for an explanation of abbreviations.

that two separate olfactory systems exist that may subserve distinct functions.

In contrast to those of the main bulb, the projection of the accessory bulb is directly into the limbic brain and terminates in the ipsilateral corticomedial nuclei (C and M) of the amygdala (Fig. 1). From the amygdaloid complex, fibers run in the stria terminalis to the bed nucleus of the stria terminalis (BNST) and the ventral amygdalofugal pathway to the medial hypothalmus (VMH) and the medial preoptic area (MPOA). Thus the accessory olfactory pathway has a fairly direct projection to those limbic structures involved in sexual and neuroendocrine regulation, but unlike the main bulb, appears not to have access to the thalamus and in turn to neocortical regions.

Functional significance of dual olfactory projections

It would appear that the evolutionary trend has been to emancipate life's events from dependence on the vomeronasal accessory olfactory system as one ascends the phylogenetic scale. Hence, reptiles such as snakes are strongly dependent on a functional vomeronasal system in order to execute behaviors such as prey detection, prey odor trailing, and several social behaviors. Male snakes with olfactory nerve section show normal courtship behavior, whereas vomeronasal nerve section prevents not only mating, but courtship and reproductively related fighting behavior among males.

Pheromonal effects on reproduction are well established in rodents, and over the past 20 years several different effects from primer pheromones have been demonstrated in female mice on exposure to male urine. These include the acceleration of puberty, induction of estrus in grouped females, and the blocking of pregnancy in newly mated females. Exposure of female mice to female urine has the converse effect of delaying puberty, inducing anestrus, and protecting against the pregnancy block. Each of these primer effects are transmitted by the vomeronasal accessory olfactory system with its projection via the amygdala to that part of the brain regulating neuroendocrine activity (dopaminergic and β-endorphin neurons of the arcuate nucleus), and their action can be universally accounted for in terms of changes in the pituitary hormones prolactin and possibly luteinizing hormone. This common neuroendocrine mechanism can account for all the reproductive effects described here, the major difference being the time of life and current reproductive status when exposure to pheromones occurs. On the basis of these findings the need to hypothesize several types of pheromone becomes redundant— all that is required is a pheromone complex signaling "maleness" and a pheromone signaling "femaleness." The olfactory block to pregnancy appears to be an exception to this simple formulation. Here "maleness" requires the further qualification of "strangeness," in that only males of a strain different from that of the original mating will block pregnancy. If the pheromone signal were indeed simply "maleness" then the block to pregnancy would be induced by all males, including the impregnating stud male. Clearly an element of strangeness is an integral part of the olfactory block to pregnancy.

The block to pregnancy by strange male odors has received considerable attention by those who adopt a sociobiological approach to reproduction. It has been suggested that the mechanism has evolved to promote heterogeneity in the population, and that strange males possessing the capacity to block pregnancy thereby increase their reproductive potential. However, such knowledge as we have of the territorial behavior and social organization of mice makes this explanation less appealing since resident males have such an advantage over intruders that access by strange males is likely to be an infrequent event. Another explanation relates pregnancy block to the effect that the pheromones have on female reproductive hormones in other contexts. Male pheromones can stimulate both early puberty and induction of estrus in grouped females, probably by suppressing prolactin secretion. Such a response, highly appropriate to these contexts, would be extremely disadvantageous following fertilization, since lowering prolactin is known to prevent implantation. Thus, some mechanism must exist to offset the general effect of the male's own pheromone on the endocrine function of his female at such times, and this recognition of the impregnating male's odor has recently been shown to be achieved by the centrifugal noradrenergic projections in the female.

In some species, such as the hamster, reproductive behavior is dependent on a functional vomeronasal system, and 40% of males with sectioned vomeronasal nerves show severe deficits in sexual behavior, while lesions to the main olfactory system are without such effects. Combined lesions of both systems completely eliminate mounting behavior in both hamsters and mice. As far as sexual behavior is concerned, there is some suggestion of an interaction between the two olfactory systems, the vomeronasal system being particularly important in naive animals for the acquisition of behavioral competence.

In higher primates and humans, the vomeronasal organ and accessory olfactory system are vestigial and only seen clearly in the developing fetus. Pheromonal influences on neuroendocrine function or behavior are minimal in these species, and no compounds exist that produce as powerful an attraction as can be demonstrated in insects. This is not because such compounds have yet to be discovered, but because primate behavior is less capable of being stereotypically released. Indeed, most mammals use all their senses to assess their natural social environment. While some species clearly rely on olfactory information more than others, primates in particular do not show fixed action patterns of behavior in response to olfactory cues. Primates, on the other hand, have all their senses well developed and, with the evolutionary enlargement of the neocortex, have the capacity to rapidly assimilate and integrate information from a number of sensory channels simultaneously. More pertinently, primates possess the ability to attend to whichever sensory channel is most relevant at the time and behavior does not come under the obligatory domination of any one sense. In other words, they use their senses intelligently to derive maximal information from the environment.

Further reading

Breipohl W, ed. (1982): *Olfaction and Endocrine Regulation*. London: IRC Press

Ritter FJ, ed (1979): *Chemical Ecology: Odor Communication in Animals*. Amsterdam: Biochemical Press, Elsevier North-Holland

Stoddart DM, ed (1980): Olfaction in mammals. *Symp Zoo Soc Lond* 45. London: Academic Press

Vandenbergh JG, ed (1983): *Pheromones and Reproduction in Mammals*. New York: Academic Press

Pleasure (Sensory)

Michel Cabanac

When a stimulus excites a sensory neuron, it arouses a tridimensional sensation (Fig. 1).

The first dimension is qualitative, identifying the nature of the stimulus; the second is quantitative, describing the intensity of the stimulus; and the third, which may be absent, is affective. The affective part of a sensation is the amount of pleasure or displeasure aroused by the stimulus. According to Young (1959), this dimension of sensation is a continuum from extreme negative affectivity (distress) to extreme positive affectivity (delight) (Fig. 1a), with indifference in the middle. A semantic indication of this continuum is implicit in the word pleasure, which turns into its antonym displeasure by the simple addition of a prefix.

Not all stimuli evoke pleasure or displeasure. In the vast, permanent flux of inputs from the sensors to the central nervous system, the large majority elicits an indifferent sensation. For example, the sight of most objects is neither pleasurable nor displeasurable. If affectivity is involved, a sense of esthetics is the source. In addition, as stated by Pfaffmann (1960), "there is almost no limit to the range of previously neutral stimuli that, by one method or another, can be made pleasurable or unpleasurable." Pfaffmann designated as primary reinforcers all stimuli creating sensations of pleasure or displeasure. The majority of them are negative reinforcers. A stimulus can be unpleasant by its very nature (e.g., a bitter taste) or by its intensity (e.g., a violent sound). When the intensity of a neutral stimulus increases, the sensation usually becomes unpleasant, as Wundt (1874) proposed. Thus, pleasant stimuli are a minority, and it is striking to observe that the range of pleasantness is limited both qualitatively and quantitatively. The most effective, and probably only, primary positive reinforcers are chemical, thermal, and mechanical stimuli.

There is an obvious relation between the affective part of sensation and behavior; the strength of the motivation for or against a stimulus is a function of the intensity of the pleasure or displeasure elicited by the stimulus. Pleasant sensations induce approach or consummatory behaviors (or both) for alimentary, sexual, and thermal stimuli. Relations exist between pleasure and usefulness and between displeasure and harm or danger. The evidence supports the hypothesis that sensory pleasure is a sign of usefulness and displeasure a warning sign. The pleasure or displeasure aroused by a thermal stimulus can be predicted from the various body temperatures of the person stimulated. Pleasure is actually observable only in transition, when the stimulus aids the subject to return to normothermia. As soon as the subject returns to normothermia, all stimuli lose their strong pleasure component and tend to be indifferent or unpleasant. This scarcity of pleasure may be more apparent than real because temperature regulation is never achieved, and normothermia is an almost virtual situation. Sensory pleasure and displeasure thus appear especially well suited as motivation for thermoregulatory behavior. The case of pleasurable flavors shows an identical pattern. Alimentary flavors are pleasurable during hunger and become unpleasant or indifferent during satiety. Measurement of human ingestive behavior confirms this relationship. Preference shows a qualitative influence: human subjects ingest more of what they like. It also shows a quantitative influence: the amount eaten is a function of the alimentary restrictions and increases after dieting, when negative alliesthesia has disappeared. The result is that pleasure scales can be used to judge the acceptability of food.

When placed in a conflict of biological motivations, e.g., fatigue versus thermal comfort, human subjects tend to maximize the sum of their sensory pleasures. By so doing they optimize their physiological functions and solve the problem of conflicting motivations.

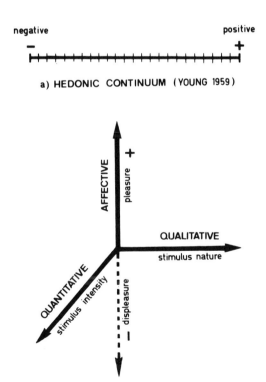

negative   **positive**

a) HEDONIC CONTINUUM (YOUNG 1959)

AFFECTIVE   pleasure   +

QUALITATIVE   stimulus nature

QUANTITATIVE   stimulus intensity   displeasure

b) TRIDIMENSIONAL SENSATION

Figure 1. The hedonic continuum. In a, as described by Young (1959), to illustrate how pleasure and displeasure merge insensibly into one another; in b, as the affective dimension makes sensation tridimensional. (From Cabanac, 1979).

Thus, the seeking of pleasure and the avoidance of displeasure lead to homeostatic behaviors. Pleasure therefore indicates a useful stimulus and simultaneously motivates the subject to use it. Both a reward and a motivation, pleasure leads to optimization of life.

Further reading

Cabanac M (1979): Sensory pleasure. *Q. Rev Biol* 54:1–29

Pfaffmann C (1960): The pleasures of sensation. *Psychol Rev* 67:253–268

Young PT (1959): The role of affective process in learning and motivation. *Psychol Rev* 66:104–125

Proprioceptive Afferent Information and Movement Control

Jerome N. Sanes

At least since the advent of motor psychophysics in the 1890s, there has been considerable interest in the role of proprioceptive information in motor control, and whether different classes of movements use such information in fundamentally distinct ways. In his classic 1899 monograph, Woodworth had some difficulty distinguishing between the contribution of exteroceptive and proprioceptive inputs to motor control. He noted that "any sense whatever may conceivably serve as the sensory basis for controlling the extent of a movement." However, it is apparent from Woodworth's discussions that muscular sense is one of the more important senses that humans use for movement control. Despite a wealth of clinical and experimental observations indicating the general significance of proprioceptors (i.e., muscle spindles, Golgi tendon organs, and tactile afferents) in motor performance, there remains some doubt about the exact contribution of these receptors to movement. That is, do certain classes of afferents contribute special regulation to certain movements?

A number of studies aimed at elucidating the role of proprioceptive inputs in voluntary movement have involved the use of relatively large mechanical disturbances imposed in the course of relatively large movements. These studies have shown that deafferented animals (surgically) and humans (functionally) are unimpaired in the performance of learned movements. There are data contrary to these results. One possible explanation for this discrepancy is that both sets of investigators are correct and that they have been examining divergent ends of a continuum in which proprioception and motor commands interact. There is some reason to believe this simply on the basis of the physiological characteristics of the input and output elements for sensory and motor control. That is, both the afferents to and efferents from the spinal cord have nonlinear response profiles, such that there are greater discharge response gains for stimuli or forces closer to threshold than for signals significantly above threshold. Although the motor control system receives proportionately more information at low stimulus levels, seemingly for greater control, the motor units are also more susceptible to modulation at force levels that result in larger changes in proprioceptive input. Thus, it may be easier to uncalibrate performance when subjects make small movements in comparison to large movements. There are data that support such a viewpoint. For example, in a comparison of the sensitivity of small and large movements to brief kinesthetic disturbances occurring during movement, Sanes and Evarts found in 1983 that only performance of the small movements was inaccurate. In these studies, the results for perturbed large movements were compatible with the notion that central motor commands can control movement adequately, but it was also clear that accuracy of smaller movements was affected by kinesthetic inputs.

Perhaps a more precise evaluation of the effects of kinesthetic inputs on motor control is the sensitivity of motor unit discharge rates to sensory disturbances. Sanes and Evarts observed that the discharge of motor units was substantially more responsive to passively imposed changes in finger position when subjects attempted to control fine changes in muscle tension so as to generate a single spike in the motor unit under observation than when subjects maintained a tension slightly above the torque required for a sustained rate of motor unit discharge. These results have a striking parallel to the results of Fromm and Evarts on pyramidal tract neurons in the motor cortex of the monkey. Identical kinesthetic stimuli occurring in association with performance of large and small movements had very different effects, with the stimuli causing intense pyramidal tract neuron discharge during small movements but having relatively little effect during large movements.

Another dimension of movement performance to consider when evaluating the role of kinesthetic inputs in motor control is the rapidity of the intended muscle contraction. Woodworth in 1899 argued that the initial phases of a movement are independent of sensory control, whereas the end of movement is under a current control, or a continuous monitoring and modification by sensory inputs. More recent research has supported the general aspects of Woodworth's hypotheses. It is well known that rapid movements or isometric muscle contractions are accompanied by a stereotyped triphasic pattern of muscle activity. Prior to actively initiated limb displacements, the agonist muscle is active briefly (A1). After the limb begins to move (but sometimes slightly before), a burst of muscle activity occurs in the antagonist muscle (ANT), while the agonist is silenced, and then a second burst is observed in the agonist (A2). The central origin of A1 is clear—it occurs before any possibility of new sensory inputs from the limb. The central origin of ANT and A2 have been established from two lines of evidence. First, patients with a large-fiber sensory neuropathy, and without clinically defined motor impairment, or subjects who have been functionally deafferented exhibited the triphasic pattern during rapid movements. Second, when humans perform rapid isometric force pulses without limb displacement (that is increases in torque to a predetermined level and then a quick release of tension), a muscle activity pattern similar to the triphasic pattern is observed. Since the triphasic pattern was never observed when active torque did not reverse direction, the triphasic pattern appears to be a fundamental property of a central motor pattern generator when intended torque rises and falls quickly after peak torque is achieved. The triphasic pattern of muscle activity occurs when torque rises to a peak in times less than about 125 msec; for longer times one observes a gradual disintegration of this pattern into a nearly pure and prolonged agonist muscle contraction.

Aside from the patterns of muscle activity that accompany voluntary movement, it is of some interest to determine whether reflexes vary according to the speed of movement.

Desmedt and Godaux have shown that as movement velocity increases the sensitivity to afferent inputs appears to fall. In these experiments subjects performed rapid or ramp movements, and a load change occurred, at the same point for both movements, in the course of the voluntary movement. During the slower movement Desmedt and Godaux observed short-latency reflex responses; these responses were absent or small during the large movement. It may be that this reflex pathway is selectively enhanced or suppressed according to the strategy of the subject.

A final topic to consider is the manual motor control of patients with a selective large-fiber sensory neuropathy. These patients are clinically characterized by profound sensory deficits, including absence of position and vibration sense, moderate decrease in pinprick, temperature, and light-touch sensation, and absence of deep-tendon reflexes. The muscular strength was normal or nearly normal in these patients. Detailed analysis of hand movements showed that with visual guidance movements and postures were relatively unimpaired, but without visual guidance patients exhibited drift from intended levels of steady state muscle activity and gross inaccuracies of movement control. The difficulties were apparent for isotonic and isometric muscle contractions, as well as single step or repetitive movements. Interestingly, although all movements were disturbed, the patients were proportionately more inaccurate when performing small in contrast to large movements. In addition, slow, alternating movements performed by the patients were accompanied by a normal reciprocal pattern of muscle activity. However, when the amplitude or frequency of movement was increased, a more cocontractive pattern appeared, suggesting that patients were unable to correctly program reciprocal activation of antagonist muscle groups for all movement types. A further point about muscular control was the observation that these patients typically (and unknowingly) showed cocontraction of antagonist muscle pairs in situations where no cocontraction was required. Thus, these patients were unable to set the appropriate levels of muscle activity necessary to maintain a stable posture or to move from one place to the next; that is, their central commands for movement and end-point control required kinesthetic inputs.

In summary, it is clear that many types of movements require some kind of kinesthetic information to mediate accurate performance. Several experiments have shown that accuracy of small movements is disrupted by disturbances, pathological or mechanical, in sensory inputs. This also seems to be true for movements that are performed slowly. Although patients with a large-fiber sensory neuropathy perform even large or rapid movements abnormally, in normal subjects these movements, at least in their initial phases, are generated by centrally programmed patterns of muscle activity. However, even large movements can be influenced by kinesthetic inputs, and their ultimate control may also depend on proprioception.

Further reading

Desmedt JE, Godaux E (1978): Ballistic skilled movements: load compensation and patterning of the motor commands. In: *Cerebral Motor Control in Man: Long Loop Mechanisms*, Desmedt JE, ed. *Prog Clin Neurophysiol* Basel: Karger 4:22–55

Rothwell JC, Traub MM, Day BL, Obeso JA, Thomas PK, Marsden CD (1982): Manual motor performance in a deafferented man. *Brain* 105:515–542

Sanes JN, Evarts EV (1983): The regulatory role of proprioceptive input in motor control of phasic or maintained voluntary contractions in man. In: *Motor Control Mechanisms in Health and Disease*, Desmedt JE, ed. *Adv Neurol* New York: Raven Press 39:47–59

Sanes JN, Jennings VA (1984): Centrally programmed patterns of muscle activity in voluntary motor behavior of humans. *Exp Brain Res* 54:23–32

Sanes JN, Mauritz, K-H, Dalakas MC, Evarts EV (1985): Motor control in humans with large-fiber sensory neuropathy. *Human Neurobiol* 4:101–114

Woodworth RS (1899): The accuracy of voluntary movement. *Psychol Rev* 3:1–114

Psychoacoustics

Joel D. Knispel

Psychoacoustics, the study of the perception of sound and music, is having a wide impact on fields of basic research such as neuropsychology and cognitive psychology, and is also contributing to practical applications in music, particularly computer music and audio reproduction.

Historically, psychoacoustics was among the first endeavors of the new experimental psychology of the 19th century. This was a consequence of the development of precise measuring instruments and a certain curiosity about the transduction of sound from physical energy into nervous/mental energy. Early psychophysical determinations of this nature indicated that the mind's representation of sound is far from linear. Sensations of sound intensity (loudness), for example, are compressed, enabling us to be sensitive to an expanded range of intensities, if not small differences in intensity. This transduction process is described by a logarithmic transformation, and the design of the modern decibel meter, which measures sound intensity, incorporates the transformation.

We are more sensitive, by more than a factor of 30, to changes in frequency, the sensation of which is known as pitch. The just noticeable difference for a tone of 2,000 Hz (cycles per second) has been shown to be 6 Hz. For a tone of 4,000 Hz it is 12 Hz. A doubling of frequency results in a doubling of the just noticeable difference. This ratio relationship also applies to the organization of musical pitch intervals (Fig. 1). The most common pitch interval is the octave. A doubling of frequency results in a tone having an identifiably similar quality to that of the first. The octave is recognized by most individuals and is found in the music of many cultures. There is some evidence that it is endogenous to the physiology of the auditory system from recordings of neurons in cat cerebral cortex that respond equally well to frequencies an octave apart but not to intervening frequencies.

Moving up functionally from the basic features of sound intensity and frequency, the spatial localization of sound depends upon ears and brain performing an intricate analysis of binaural cues and shifts in spectral sensitivity. Localization in the horizontal plane (azimuth) occurs through a convergence of sensory input from both ears onto single neurons beginning at the level of the superior olivary complex of the brain stem. For frequencies of less than 1,500 Hz, differences in time of arrival of a sound at both ears are detected by these cells. Frequencies above this exceed the resolving power of the cells. At frequencies above 1,500 Hz, interaural differences in intensity provide the localization cues. In order to detect whether a sound is coming from directly in front or behind, we make use of the architecture of the outer ear, which acts as a selective filter. Sounds being emitted from in front are transmitted to the middle and inner ear with a relatively linear frequency response at moderate intensity levels. The frequency response to sounds from the rear, as well as from the sides, is distorted to provide a shift in spectral sensitivity. The same holds true for localization in the vertical plane. For example, as the sound source rises there is a characteristic peaking of the response at 9 kHz. Computer music composers can provide a convincing illusion of sound rising by gradually increasing the energy in the 9 kHz band.

Acoustic sounds usually consist of more than a single frequency. Striking one key on the piano produces a harmonic spectrum of frequencies (partials) that includes the fundamental frequency, which defines the note, and multiples of the fundamental that convey the characteristic tone quality of the instrument. This is known as timbre. The same note played on the clarinet has a similar fundamental frequency but different overtone structure that distinguishes it from the sound of the piano (Fig. 2).

In concert, the sounds of the instruments maintain their separate identities through a phenomenon known as spectral fusion. The spectrum of one does not bleed into the others'; the sounds of the instruments are perceived as separate entities through several mechanisms including template matching to memory, onset asynchrony, and shimmer or vibrato. The last two represent forms of frequency modulation. Vibrato is periodic. Shimmer is aperiodic. Many instruments, particularly voice or blown instruments, do not maintain a steady pitch but vary around a center frequency. Computer music programs can include a factor for variation around a center frequency of up to 2% to impart a life-like quality. These variations are time-locked across the entire harmonic spectrum of an instrument so that the fundamental and overtones vary together, in effect fusing the spectrum and distinguishing it from those of other instruments. Computer simulations of these mechanisms suggest a model for auditory attention, or how we are able to pick a voice out of the crowd. First, sounds with spectral peaks indicative of voice are synthesized. These alone are not sufficiently reminiscent of voice. When vibrato is superimposed, however, a voice emerges. When several asynchro-

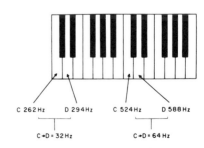

Figure 1. A section of the piano keyboard showing representative approximate pitch intervals in Hz. The interval between C notes is an octave described by the frequency ratio 2:1. The absolute frequency difference of all pitch relationships is doubled from one octave to the next.

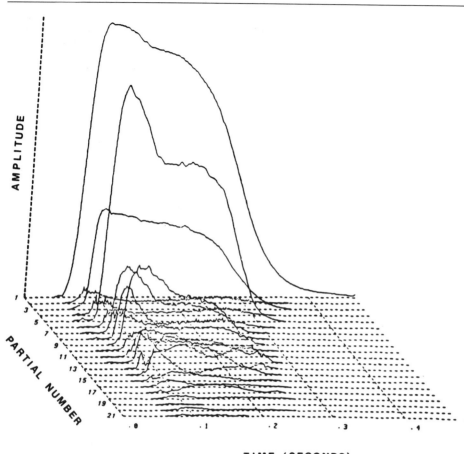

AMPLITUDE

PARTIAL NUMBER

TIME (SECONDS)

Figure 2. Perspective plot of amplitude × harmonic number × time for a clarinet tone played at E-flat, approximately 311 Hz. The fundamental is plotted in the background with higher frequency overtones in the foreground. From Morrer JA, Grey J (1977): Lexicon of analyzed tones. *Computer Music J* 1:12–29. Reprinted with permission.

nous vibratos are superimposed on different spectral peaks, several voices can be identified and focused upon.

Much is known about the neural mechanisms underlying loudness, pitch, and localization. Less is known about the mechanisms subserving music. From clinical case histories and neuropsychological studies, there appear to be functional distinctions between the cerebral hemispheres. Individuals suffering cerebral lesions can show a variety of musical deficits depending upon their premorbid ability and the site of injury. Amusia is rare. It is typically, but not always, associated with aphasia (loss of language) and agnosia (loss of meaning) for nonverbal sounds. In amusia, the lesion is often bilateral, in the area of the temporal lobe. When the injury is unilateral, it is predominantly found in the right hemisphere. In general, the left cerebral hemisphere involves verbal and associative functions, and the right involves perceptual and discriminative functions. The right hemisphere is dominant for melody perception in nonmusicians. In musicians, the left hemisphere is dominant. It is tempting to speculate that musicians who have learned the language of music transfer this function to the verbal hemisphere.

Prospective developments

New noninvasive techniques are used to monitor electrophysiological activity in intact functioning individuals. One such method uses an external magnometer sensor and triangulation procedures to access magnetic activity from the depths of the

cerebral cortex. As the primary auditory area is tucked into an invagination, this sensor now makes it possible to intimately study auditory activity in normal individuals. The device has already mapped the receptive fields of the area. Another technique uses multiple scalp electrodes to record auditory evoked potential responses in the electroencephalogram (EEG). Through computer graphics, motion pictures of the evolution of the response across the surface of the brain reveal a good deal of information about the cortical processing of sound.

This accumulated knowledge may soon find clinical application in behavioral medicine and psychotherapy. Methods involving feeding EEG signals to computer music synthesizers for psychological applications are currently being evaluated. Individuals connected to synthesizers in this way have been able to play specific sounds and repeatable melodic lines upon demand. It may soon be possible to develop musical instruments that would bypass the muscular system; instead music would be produced through direct interface between brain and computer. Such music also has value as a stress reduction technique. It is particularly appealing to the individual wearing the electrodes, especially if listened to in real time. Establishing an immediate feedback loop of this sort is conducive to relaxation. EEG synchrony consistent with relaxation has been seen in association with this behavioral outcome. The efficacy of computer music programs designed to enhance particular emotional and therapeutic states can be evaluated by monitoring physiological processes. The program can then be iteratively updated to achieve the desired state, as in biofeedback.

Further reading

Bodis-Wollner I, ed (1982): Evoked potentials. *Ann NY Acad Sci* 388

Hecaen H, Albert ML (1978): *Human Neuropsychology*. New York: Wiley

McAdams S (1982): Spectral fusion and the creation of auditory images. In: *Music, Mind and Brain*. Clynes M, ed. New York: Plenum Press

Warren RM (1982): *Auditory Perception*. New York: Pergamon Press

Psychophysics

Trygg Engen

G.T. Fechner originated psychophysics in the 1860s to describe mathematically the relationship between body and mind, the conscious experience of a sensation resulting from an external physical stimulus. Psychophysics had an important immediate impact on psychology, sensory physiology, and related fields, because it provided a means of measuring sensation which previously, like all other aspects of the mind, had been considered private and immeasurable. Knowledge of such variables is essential in human factors research on the compatibility of the environment and equipment with human sensory ability, and in evaluating human error. Historically, psychophysics promised to make possible a quantitative and scientific study of what was then described as "higher mental processes," now a topic in cognitive science. It had a direct effect on Binet, for example, who developed the now universally used methods for measuring intelligence.

Although Fechner's interest in this problem was largely philosophical, he was an accomplished physicist, familiar with the use of mathematics, and wanted to apply ideas from these fields to mental measurement and thus make it scientific rather than philosophical. He reasoned that if one knows the mathematical form of the psychophysical relation between a physical variable and its corresponding sensation (the "psycho" term of the relation), one can measure mental attributes by measuring their physical correlates.

Research has shown that psychophysical relations, for example, between sound pressure and sensed loudness, are rarely linear. Typically, the intensity of a sensation increases as a negatively accelerated function of equal physical increments in physical intensity. For example, starting at the low end in adjusting the loudness of a sound source and increasing it by equal increments results in the sensation of smaller and smaller increments. The sensation of pain associated with electric shock and the sensation of heaviness of weights are atypical in that equal increments of these physical variables are sensed as larger and larger.

Psychophysical scaling

The most direct method for obtaining a psychophysical relation is to ask observers straightforwardly to make quantitative judgments of sensed magnitudes corresponding to physical stimuli. The observer might assign a number to each stimulus; the higher the number the greater the sensed magnitude. Other methods ask the observer to match the sensed magnitude of one stimulus to that of another physical stimulus. For example, the observer might match the loudness of a tone by adjusting the length of a line. In still another method, the observer is asked to produce certain prescribed ratios of sensations such that one stimulus produces twice the sensory magnitude of another stimulus. Such psychophysical judging and matching is considered similar to measuring and calibrating in other scientific domains (Stevens, 1975).

As an illustration, in one experiment observers were asked to select a weight which seemed half as heavy as another weight (Engen, 1971). These target weights were 150, 300, 550, or 900 grams. The average half-heaviness judgments from a group of observers are presented in Table 1.

The first column presents the weights of the target weights, the second column the average judgments in grams. The third column shows the ratios between the weight of the standard and the weight judged half as heavy. The required half-heaviness criterion of 2:1 is listed in the fourth column. The results illustrate that the psychophysical relation is nonlinear (compare columns three and four), for the obtained physical ratio of the weights are smaller than the subjectively experienced ratios without exception. There is also a certain numerical constancy in the ratios, indicating the nature of the mathematical form of the psychophysical relation. The physical ratio is always about 1.7 when the experienced sensory ratio is 2.0, indicating that sensed heaviness changes faster than physical change in weight; increasing a weight by a factor of 1.7 will make it appear to be twice as heavy. Thus, in this case, sensed magnitude grows faster than the physical. Mathematically, the general rule is that equal physical ratios along a stimulus dimension will be experienced as producing equal experienced ratios of sensations.

But it is important to bear in mind that the size of a physical ratio, for example, one eliciting a subjective ratio of 2:1, vary for different stimuli. To compare with the heaviness data above, as the intensity of a sound is increased by 10 dB (SPL), or 10 times in terms of pressure, the sensed or subjective increase in loudness is only doubled. This holds whether one changes from 50 to 60 or 100 to 110 dB. Consequently, if one scales both the increase in subjective loudness as a function of pressure and the sensations experienced in logarithmic values (dB is already a logarithmic value), to indicate addition of magnitude on each scale rather than ratios, one obtains a function of the form $y = ax + b$, namely

$$\text{Log } \psi = n \log \phi + \log c \qquad (1)$$

where ψ represents sensation, ϕ physical intensity of the corresponding stimulus, n, the exponent or slope of the function,

Table 1. Ratios of Weights versus Judged Heaviness

Standard in Grams	Half-Heaviness Judgment in Grams	Physical Ratio	Subjective Ratio
150	94	1.60	2.00
300	159	1.89	2.00
550	325	1.69	2.00
900	542	1.66	2.00

and c the intercept which depends on methodological factors we need not consider here (see Engen, 1971). Expressed in linear terms the function becomes $\psi = c\phi^n$. The most interesting constant is n which indicates how sensation increases or decreases as physical stimulation varies.

It is called a power function because it states that sensory magnitude is proportionate to physical stimulus magnitude raised to a power. It is sometimes called Stevens' law, because of his outstanding work in demonstrating its general validity. The power is represented by the exponent (n) of the function which describes the effect of different stimuli, as illustrated with the weights and sound pressures. It has been observed to be as low as 0.10 for odor intensity under certain conditions, close to 1.0 for length of lines, and as high as 3.5 for the intensity of electric shock to the fingers. It must be noted that the power function is not perfectly reliable. Both the intercept (c) and the exponent may vary with the range of stimuli presented, the state of adaptation, and other characteristics of the individual observer (Engen, 1971).

In general all researchers do not accept the power function as the only possibly valid one. Based on a very different approach, Fechner himself proposed a logarithmic function,

$$\psi = k \log \phi \qquad (2)$$

where ψ and ϕ are as above, and k a constant determined by the stimulus threshold, representing the lowest physical value eliciting a sensation, and differential threshold providing a subjective unit of sensory intensity. The most important point is the proposal that sensation increases in arithmetic steps as the physical stimulus is increased in logarithmic steps. Fechner got the idea of this logarithmic function from results with methods assessing the subjects' ability to detect small differences in stimuli.

Stimulus detection

A classical psychophysical assumption is that there are stimulus thresholds. The essential aspect of the Fechnerian approach is to measure them, to divide stimuli (or stimulus differences) into those which the subject can detect and those he or she cannot. It is assumed that although this threshold will vary for different individuals, at any one moment in time it is fixed for any one individual. It is in this sense that a threshold may be referred to as absolute, referring then to the weakest stimulus a person can detect. Presumably, it varies from moment to moment because of spontaneous activity in the nervous system, the efficiency of the connecting neural pathways, etc. It is assumed that this variation is random and that the sensitivity of a sense modality may be described in terms of the normal distributions fitting other biological systems.

The absolute threshold, the lower limit on the physical dimension, is still described as RL, the abbreviation for the German *Reitz Limen*. The method of limits is the prototype for its measurement. Stimuli from below to above threshold are presented to the observer in small steps, ordered according to physical size, and presented in ascending and descending series in order to determine the limit in the series where the observer can detect one but not the adjacent one. In determining an audiogram, for example, a tone of a certain frequency is presented at various decibel levels of intensity. The threshold is the average decibel level, obtained from many ascending and descending series of stimulus presentations, dividing the levels into those which the observer can hear and those he or she cannot.

While RL refers to the border between detectable and nondetectable stimuli, the difference limen or threshold (DL) refers to the minimum stimulus change required for the observer to detect that there has been a change in the level of stimulation. For example, if an observer is listening to a 1000 Hz tone, the question is how much change in frequency (Hz) will be required before he or she will hear a change in pitch. This sensory ability is typically measured with the method of constant stimuli. It uses about seven predetermined discrete stimuli, again physically ordered from low to high. The middle stimulus of the series is designated as the standard (St), and it is compared in a random order with each of the others, designated as comparison stimuli. The task is to report for each pair whether or not a difference can be detected. The stimuli cover only a very narrow range with a small step between adjacent stimuli and thus present a difficult decision task where perfect performance is not possible. The DL is defined as the physical difference from the standard (either smaller or larger) the observer can detect 50% of the time, based on a number of trials. This statistically defined physical difference is the DL, also known as the just noticeable difference (jnd), and δS. For example, one might obtain a DL of 3 Hz with a 1000-Hz standard with comparison stimuli ranging from 994 to 1006 Hz in 2-Hz steps.

The Weber fraction

Fechner applied such discrimination data in developing the logarithmic function. He calculated the so-called Weber fraction or constant, named for E.H. Weber, a physiologist who studied cutaneous sensitivity. He discovered that the ability to detect changes in touch from weight placed on the skin was not absolute, requiring a constant physical change, but relative to the pressure already exerted on the skin as defined by the standard stimulus. Accordingly, the ratio of the DL to the standard or St will be a constant (k), or

$$k = DL/St. \qquad (3)$$

In our example, these values are 3/1000 = 0.003 or 0.3%. Thus, if one was to do the same experiment with a higher St of 3000 Hz, DL should equal 0.003 times 3000, or 9 Hz, as 9/3000 = 0.003. In general the DL increases logarithmically with linear increases in stimulus intensity.

Fechner and followers assume that the subjective difference experienced is the same for 3 Hz for a 1000-Hz standard as for 9 Hz obtained with a 3000-Hz standard, and so on. The DL can then be used as the unit in the logarithmic psychophysical function.

Contemporary concerns

The main support for the logarithmic function has come from electrophysiological work, such as the frequency of impulses obtained from an electrode in the visual system of the horseshoe crab stimulated by light intensities, and the summated neural response from the chorda tympani of the human tongue stimulated by different concentrations of a tastant. By comparison, the main support for the power function has come from tests with quantitative judgments by intact human observers. However, the results are not unequivocal. A related dilemma is that each function is based on an assumption which itself is not directly observable, and therefore conclusions about the validity of each function can only be based on indirect evidence. For the direct methods supporting the power function, the problem is the assumption that the subject is capable of describing experiences quantitatively as though he or she were applying a subjective meter stick. For the logarithmic function, the problem is the assumption that a subjective unit may be derived from judgments which are nonmetric. There is not enough information for any final conclusion (see Baird and Noma, 1978).

Regardless of the validity of the assumption that the DL provides a sensory unit, the Weber fraction has been found to be a practically useful rule of thumb for sensory resolving power. The exception is for very weak stimuli which are not always detected or very strong ones which may involve a change in sensory quality. For example, as sound pressures reach high levels one may feel pain in addition to hearing sound. In such cases, Weber's fraction is likely to change.

Fechner and Stevens had in common that they put the emphasis on the stimulus and the importance of defining it in order to understand the response, applying what might be characterized as a stimulus-response (S-R) model. Recently, there has been an increasing tendency to emphasize the active processing of stimulation by the observer (O) in what might be described as S-O-R models. Thus, so-called detection theory argues that people interpret incoming stimulus information in terms of past experiences and present motivation. They are not passive recipients. A soldier observing a radar screen is a different judge in war than in peace.

Theoretically, debate about the validity of sensory measurement has continued without stop since Fechner proposed the logarithmic function (see Baird and Noma, 1978). However, a distinction was made between psychophysics as theory and methodology from the very beginning. Primarily, psychophysics provides the means for measuring the sensory reactions to physical stimuli. The classic methods have been modified and new ones added, such as the direct scaling methods, to deal with specific situations. The methods also include master scaling to handle direct scaling of stimuli by different individuals in different situations. And animal psychophysics is a field which developed to deal with various psychophysiological problems for which one cannot use human subjects. There has thus been a constant expansion of psychophysics. While it was originally part of the domain of sensory psychology and physiology, it is now a technology of general application.

Finally, progress has been made in psychophysics by moving away from two traditional preoccupations (Engen, 1984). One is the overemphasis on the absolute threshold and sensitivity. Psychophysics now has a more balanced approach with due attention to weak stimuli as well as more effective suprathreshold stimuli. The other was the preoccupation with "pure" stimuli, such as tones and basic tastes, which had its origin in Muller's doctrine of specific energies of nerves and the "mental chemistry" it fostered. Research has shown that speech sounds, for example, are not related to pure tones in any simple manner, and as a result psychophysics has been directed more and more toward more complex and presumably significant stimuli, such as speech sounds and food odors.

Further reading

Baird JC, Noma E (1978): *Fundamentals of Scaling and Psychophysics*. New York: Wiley & Sons

Engen T (1984): Classical psychophysics: Humans as sensors. In: *Clinical Measurement of Taste and Smell*, Meiselman HL, Rivlin RS, eds. Lexington, Mass: Collamore Press

Engen T (1971): Psychophysics I. Discrimination and detection and II. Scaling. In: *Woodworth and Schlosberg's Experimental Psychology*, Kling JW, Riggs LA, eds., 3rd ed. New York: Holt, Rinehart and Winston

Stevens SS (1975): *Psychophysics*. New York: Wiley & Sons

Psychophysics and Neurophysiology

Donald M. MacKay

Psychophysics can be broadly defined as the quantification of sensory experience. This entails not only the assessment of human powers of signal detection and sensory discrimination but also the calibration of subjectively perceived intensities and other parameters of stimulation.

It was for some time thought obvious that the data of psychophysics should be directly and quantitatively comparable with those of sensory neurophysiology. In the classic case of color vision, for example, the correspondence between photometric curves for sensory receptors and normal psychophysical data is dramatically close. Again, for many years physiologists felt that their finding of a (roughly) logarithmic relation between physical stimulus intensity and receptor ganglion cell firing rates was a direct confirmation of Fechner's psychophysical law that the subjective scale of perceived intensity is logarithmically related to physical intensity.

More recent evidence has thrown doubt on the presupposition of direct comparability. S.S. Stevens developed improved and self-consistent methods of estimation that showed subjectively perceived intensity to be a power function of physical stimulus strength, but efforts to demonstrate that the neurophysiological data could also fit a power law met with only patchy success. Further difficulties were encountered when efforts were made to account for quantitative perceptual properties such as the size constancy or stability of the visual world by theories presupposing that they demanded directly corresponding properties in some neural image derived from the retinal input.

An alternative approach that avoids such awkward assumptions postulates that the direct neurophysiological correlate of psychophysical properties is to be sought, not in the incoming sensory stream or its derivatives, but in the matching changes elicited in the organizing system that is presumed to keep up to date the organism's conditional readiness to reckon with the perceived world. In this view, when a subject estimates sensory intensity, the neural correlate of his perceptual experience would be an internally generated matching response, automatically adjusted so that (in at least some respects) its internal effects are equal and opposite to the internal disturbance produced by the sensory input. The perceived intensity of the stimulus might plausibly be related to the physical intensity of the internal effort required to generate the matching response, rather than to the sensory input itself. Suppose, for example, that the matching response generator had a logarithmic characteristic like that of the sensory transducer, so that its output $f_m = k_m \log \psi$, where ψ represents the internal effort; and that the sensory transducer output $f_s = k_s \log I$, where I represents the physical intensity of stimulation. Then when $f_m = f_s$ we would have $k_m \log \psi = k_s \log I$, or $\psi \propto I^\beta$, where $\beta = k_s/k_m$. Thus if we adopt a matching response model, the assumption of logarithmic transfer characteristics actually predicts a power-law relationship between the internal efforts

elicited by a stimulus and its physical intensity, showing that there is no necessary conflict between Stevens's power law and either Fechner's or the neurophysiologists' logarithmic functions.

Perceptual illusions as clues to sensory processing mechanisms

A second area in which psychophysics challenges neurophysiology is the study of perceptual illusions; here again the temptation to confuse perceptual and physiological categories can be a trap, and these phenomena serve better as experimental suggestions than as predictors of physiological findings. As long as sensory mechanisms are functioning effectively, study of our perceptual experience may no more elucidate the principles on which they operate than the viewing of a TV set reveals the mechanisms it employs. But in either case, if we can find input signals that give rise to distorted or spurious performance, the nature of the malfunctions and the conditions giving rise to them can provide valuable clues to the processes underlying normal functioning. Studied in this light, perceptual illusions can be of considerable neuroscientific value. If, however, perception is mediated by an internal matching response, it would be unsafe to regard all perceptual illusions as evidence of physiological malfunction on the side of the sensory input, since distortions in the generation and processing of the matching response could equally be to blame.

Obvious examples of physiologically relevant illusions are negative visual afterimages of color brightness, whose time-course can be reasonably correlated with that of biophysical and biochemical changes in retinal sensitivity. A widely used trick is to expose only one eye to the adapting stimulus, and then see how well the illusion can be seen when the other eye is used. Unfortunately, as a way of distinguishing retinal from central effects this method has snags. The adapted eye, although closed, may continue to contribute to the signal stream.

It is tempting to suppose that a negative perceptual aftereffect of exposure to a specific stimulus betokens physiological fatigue of a corresponding channel, which could lead to a shift in the zero-level of the signals transmitted. Although this idea suggests experimental questions, it is unsafe as a general predictor. The familiar waterfall aftereffect of retinal image motion, for example, suggests an interesting question, whether cortical cells specific for motion in one direction show physiological signs of adaptation after prolonged exposure. What it cannot logically do is to predict with any confidence the time course or magnitude of such adaptation.

Perhaps the most striking cautionary example is the McCollough aftereffect, whereby prolonged exposure to alternating red and green gratings at orthogonal orientations makes a black-and-white test grating appear tinted with the hue complementary to that originally associated with its orientation. When first described it was widely taken as perceptual evidence of fatigue in cortical cells selectively sensitive to both the orientation and the color of bars or edges of luminance falling on their retinal receptive fields. As such, of course, it could be expected to decay most rapidly under conditions favoring recovery from physiological fatigue, such as the removal of excitatory inputs. In fact, however, when one eye was kept in complete darkness for 25 hours after McCollough exposure and then tested, the chromatic aftereffect was still at full strength, although the effect in the other eye had decayed in the normal way to less than 10%. Rather than betokening physiological cell fatigue, then, this negative aftereffect suggests some form of compensatory associative change, possibly in synaptic couplings between neurons, which remains undisturbed until fresh input signals break up the associations. The fact that the two eye channels behave relatively (though not totally) independently suggests that the neural networks responsible are in the uniocular visual pathways.

Perceptual stability

A third example of the pitfalls besetting the physiological interpretation of psychophysical phenomena is afforded by the problem of perceptual stability. During exploratory eye movement the retinal image dances all over the receptor mosaic, and its neural image similarly dances over the primary visual cortex, yet no corresponding instability is perceived in the visual world. On the other hand, if the world is viewed through a mirror that is saccadically rotated to produce similar retinal and cortical image movements, it appears to jump about. On the presupposition that stability of perception requires stability of some neural image, physiological theories have been produced that postulate the removal or cancellation from the incoming signals of the changes produced by exploratory eye movements, under the guidance of an elaborately detailed corollary discharge or *efferenzkopie* from the oculomotor system: a formidable task of high-precision information engineering.

The presupposition underlying such theorizing is, however, logically unfounded. If perception is a matching response whereby the perceiver's organizing system is updated to reckon with the demands represented by the sensory input, then the logical requirement for stability in the world-as-perceived is simply that no sensory input should falsify the null hypothesis represented by the current state of organization. Now the changes in retinal input caused by exploratory eye movement, so far from disconfirming the current null hypothesis, are precisely what the eye movement was calculated to bring about on that null hypothesis. In short, when evaluated for their information content, they confirm rather than challenge the stability of the world-as-perceived. There is thus no need to remove them from the incoming signal stream. Instead of elaborate and precise cancellation mechanisms, all that is logically required at the neurophysiological level is a process of evaluation, under criteria specified by the system responsible for initiating ocular exploration. The corollary discharge required for this purpose need have no greater precision than the exploratory process itself. A similar analysis shows that for such perceptual phenomena as the constancy of perceived sizes of objects during relative distance changes it is logically unnecessary to postulate any zoom lens mechanism in the visual pathway.

The moral once again is that the use of psychophysical data as direct predictors of physiological observations is at present too heavily theory laden to be scientifically secure; but their value as a stimulus to experimental questions in neuroscience has already been considerable, and as basic principles are clarified, the interaction of the two is likely to become increasingly fruitful.

Further reading

Dartnall HGA, ed (1972): *Photochemistry of Vision: Handbook of Sensory Physiology*, vol 7/1, New York: Springer

MacKay DM (1970): Perception and brain function. In: *The Neurosciences: Second Study Program*, Schmitt FO, ed New York: Rockefeller University Press, pp. 303–316

MacKay DM, with MacKay V (1976): Retention of the McCollough effect in darkness: Storage or enhanced read-out? *Vis Res* 17:313–315

MacKay DM (1973): Visual stability and voluntary eye movement. In: *Handbook of Sensory Physiology*, Jung R, ed, vol 7/3a. New York: Springer

Sensory Receptors, Cutaneous

Ainsley Iggo

The skin of mammals is richly innervated and contains sensory receptors specialized for the detection of three particular categories of natural stimuli: (1) mechanical contact (tactile receptors), (2) temperature changes by contact and from radiation to or from the body surface, and (3) actually or potentially damaging traumatic and chemical insults. The wide range of energies in such a diversity of natural stimuli requires the presence of diverse receptors.

Transduction

The essential primary function of sensory receptors is to convert (or transduce) natural stimuli into the standard code by which nervous systems work—the self-propagating action potentials that travel along the afferent fibers away from the skin to the central nervous system. The processes of transduction in the receptors are not well understood. An analogous process, taken from high fidelity sound reproduction, is the recorder cartridge, containing a piezo-electric component that converts movement to a voltage change.

The central question about sensory receptors is, How do they convert natural stimuli into nerve action potentials? It is now well established that a common feature is the generation within the receptor of a flow of current, recorded as a potential change, proportional to the intensity of the applied stimulus. The current flow in the nerve terminal sets up nerve action potentials at a spike initiation site that then travel centrally along the afferent nerve fiber. A well-studied example of the transducer mechanism is the hair cell. Best known from work on lateral line organs in fish, this cell is also the essential detector in the vertebrate ear and labyrinth. The apical end of hair cells contains rod-like stereocilia, bathed in an external medium rich in potassium ions. Mechanical displacement leads to an altered permeability of the tips of these stereocilia, causing an inward flow of current at the apical end of the hair cell that passes out across the basal part of the hair cell, probably causing entry of calcium ions. These in turn lead to release of transmitter that by a change in membrane permeability generates the current in the afferent nerve terminal required to generate action potentials more proximally in the axon. This complex series of events requires as a starting point the as yet unknown membrane transducer. The hair cell is a highly specialized form of mechanoreceptor that nevertheless illustrates some of the essential features of the transducer process. Other mechanoreceptors lack the special receptor cell, and all transduction occurs in the nerve ending itself.

Mechanoreceptors

The functional specialization of cutaneous receptors is broadly paralled by structural specialization. The cutaneous sensory receptors can be broadly classified into those that have elaborate, usually encapsulated, structures that lie close to the dermoepidermal border or slightly deeper in the dermis and others with simple unencapsulated terminals. The afferent nerve fibers are all closely invested with a Schwann cell that in the larger (2–20 μm) axons forms the characteristic lamellated myelin sheath. At the nerve terminal, the investing Schwann cells may form an elaborate multilayered coat, almost completely covering the morphologically complex nerve terminal, and this may in turn be contained at the center of a structure formed from cells of dermal origin, the whole structure forming a complex sensory receptor. An example is the Pacinian corpuscle, which has a many-layered outer coat that both physically protects the nerve ending at the core and, more importantly, significantly modifies the physical stimulus that penetrates to the center and excites the transducer element of the nerve terminal. In the Pacinian corpuscle the lamellae of the outer coat act as high-pass filters so that the receptor can detect high-frequency vibrations—less than 1 μm in amplitude and at frequencies as high as 1500 Hz.

Several distinctive encapsulated sensory receptors exist in mammalian skin, and they sometimes are grouped together in special cutaneous sense organs, such as sinus hairs, especially round the mouth in many species, or the highly distinctive Eimer's organ in the snout of the mole. Specific examples of encapsulated receptors are: Pacinian corpuscles, Meissner corpuscles, Krause endings, and Ruffini endings. These different structures are all innervated by fast-conducting myelinated fibers (A or group II) and are mechanoreceptors (tactile receptors). They are joined by a second group of receptors that although not forming independent encapsulated structures nevertheless have an organized and distinctive morphology. These are the hair follicle receptors that encircle the hair shaft and the Merkel cells—Merkel discs found at the base of the epidermis. These last are always found within the basement membrane of the epidermis in association with a modified epidermal structure. All these sensory receptors are mechanoreceptors, and each has a distinctive range of properties that enable it to analyze a particular parameter of a mechanical stimulus. Thus the Pacinian corpuscles detect high frequencies of displacement (up to 1500 Hz), the Meissner corpuscles (of hairless skin) and the hair follicles (in hairy skin) detect middle range frequencies (20–200 Hz), and the Merkel cells and Ruffini endings detect steadily maintained deformation of the skin (DC to 200 Hz).

Thermoreceptors

The temperature of the skin is detected by a quite separate set of sensory receptors that are specialized to respond to both constant and fluctuating skin temperatures. This they do by possessing a sensitivity to skin temperature—not to some other parameter such as the wavelength of radiation, so they are not photodetectors like the rods and cones of the retina of the eye. At constant temperatures, the receptors can discharge impulses continuously and thus provide a long-term indication of skin temperature. They are more sensitive, however, to changes in skin temperature, and this dynamic capability may be at least as important as the static one. Two main kinds of sensitive thermoreceptor exist in mammalian skin–cold receptors with peak sensitivity around 25°–30°C, excited by dynamic downshifts in temperature, and warm receptors with peak sensitivity around 39°–40°C, with a dynamic sensitivity to rising temperatures. The static and dynamic ranges of the two sets of receptors overlap, and in the range 30°–40°C it is likely that elements of both sets are active. The nerve fibers for these receptors are thin. In primates the cold receptors are A fibers, conducting around 5–15 m/sec, whereas the warm fiber axons are nonmyelinated, conducting slowly at less than 2 m/sec. The structure of thermoreceptors is unresolved, but we know that they are not encapsulated.

Nociceptors

The third main class of cutaneous receptors are the nociceptors, so-called because they respond to noxious stimuli, such as heating the skin, or strong pressure or contact with sharp or damaging objects. The distinctive feature of these receptors is their relatively high threshold which ensures that they are more or less indifferent to those mechanical and thermal stimuli that may fully activate mechanoreceptors and thermoreceptors. The high thresholds confer on nociceptors the important property of being able to function as warning devices, thus enabling the organism to take protective action. Two principal kinds of nociceptor exist in mammalian skin. The first has myelinated axons conducting between 10 and 40 m/sec (A fibers) and receptors that are best adapted to respond to mechanical stimuli, such as pin-pricks or squeezing the skin. The receptor terminals are morphologically simple and are reported to be in the basal layers of the epidermis. The second kind respond to a diversity of stimuli, especially to high (>42°C) or low (<10°C) temperatures, and to pain-producing chemicals, as well as high-intensity mechanical stimuli. Afferent fibers are slow-conducting, either nonmyelinated or thinly myelinated (small A or C). A distinctive feature of these nociceptors is that their axons contain the neuropeptide substance P (SP), but its role is still unclear. The receptor morphology of these so-called polymodal receptors is very simple, and they do not have encapsulated receptor terminals. An interesting and important feature of nociceptors is that they have enhanced sensitivity in inflamed tissues, and may then be excited by normally innocuous stimuli. This high sensitivity can be mimicked by the simultaneous treatment of normal nociceptors

with a prostaglandin, such as PGE_2, and a peptide, such as bradykinin. Since these chemicals are known to be produced in inflamed tissues, enhanced sensitivity of nociceptors may well be due to them. Further support for this suggestion comes from the pain-relieving action of aspirin. This prevents the formation of prostaglandins and so could remove an essential potentiating agent and thus reverse the enhanced sensitivity of nociceptors.

Receptors and Sensation

Recent studies in conscious humans have thrown new light on the sensory functions of the cutaneous sensory receptors. Direct recording from, and electrical stimulation of, single nerve fibers in peripheral nerves have done two things. First, they have confirmed that human skin contains receptors that behave in the same way as that of laboratory animals, and thus greatly enriched knowledge of the human receptors, since we can now proceed by analogy with animal studies. Second, and quite dramatically, they have established that isolated activation of an individual sensory receptor can result in distinct sensory perceptions. Meissner corpuscles evoke touch, Merkel receptors evoke pressure, and nociceptors cause pain. A good correspondence between the anatomical structure of large mechanoreceptors (Pacinian corpuscles, Meissner corpuscles, Merkel cells) and elementary sensations is now emerging from this work. The full complexity of sensations and perceptions that can be aroused by the normal activation of a population of many hundreds of different cutaneous receptors, in different degree and extent, is of course much greater and dependent on interactions among the central nervous processes called into action, and should not be seen as a simple one-to-one process. Nevertheless the encoding of specific information is already begun by sensory receptors in the skin, and the central nervous system makes its further analysis using these already specified building blocks.

Further reading

Darian-Smith I (1984): The sense of touch: Performance and peripheral neural processes. In: *Handbook of Physiology, vol 3, pt 2*. Darian-Smith I, ed. Bethesda: American Physiological Society
Iggo A, Andres HK (1982): Morphology of cutaneous receptors. *Ann Rev Neurosci* 5:1–31
Iggo A, Iversen LL, Cervero F, eds (1985): *Nociception and Pain*. London: Royal Society
Hamann W, Iggo A, eds (1984): *Sensory Receptor Mechanisms*. Singapore: World Scientific
Handwerker HO, ed (1984): Nerve fibre discharges and sensations. *Human Neurobiol* 3:1–58

Sensory Transduction, Small-Signal Analysis of

John Thorson

Applied mathematics is a mainstay of quantitative physiology—from the cable theory of the cell membrane to the cross-bridge theories relating macromolecular events and tension in muscle. A subset of this work has applied engineering transfer-function analysis or input-output analysis to the dynamic responses of sensory cells of several modalities. This approach involves close control of inputs such as light and mechanical deformation, and monitoring of output changes of generator potential and spike rate. Moreover, with small sinusoidal or step-like changes of such inputs, the differential equations describing the measured input-output dynamics are often nearly linear.

The forms of such linear equations, which can reflect specific rate-limiting processes such as diffusion, transport delays, and particular chemical kinetics, can suggest testable hypotheses about the events underlying the sensory transduction. Also, the transfer functions of separable processes (e.g., transduction and inhibition) measured individually within, say, an eye can be put together mathematically to see whether the combination explains measures of integrated imaging. In the latter case, as in Hodgkin's and Huxley's 1952–53 bottom-up analysis, the focus is on the next higher level of explanation, rather than the next lower. Some progress in these directions has been achieved in the *Limulus* eye and in invertebrate mechanoreceptors, and related techniques are being applied in the vertebrate retina.

A difficulty in such small-signal analyses arises as follows: When adaptation of most receptor cells is studied over several decades of frequency or time with small fluctuations of their inputs, their responses show nearly linear dynamics under a given set of conditions. However, this linear behavior has rarely suggested hypotheses about mechanism because the differential equations required are often of fractional order—that is, step responses relax approximately according to a power law in time (t^{-k}, $0 < k < 1$, $t > 0$, k a constant), and frequency responses rise correspondingly with fractional powers of frequency (f^k, with phase leads near $k\pi/2$ radians). The second-order partial differential equations of diffusion can predict such behavior with $k = 0.5$, but measured values of k in various receptors can be 0.2, 0.3, 0.75, etc., and physical models have not been directly available. This near-linear power-law behavior (k an exponent of frequency and time) should not be confused with the nonlinear power-law descriptions of sensory systems over large ranges of input magnitude (k an exponent of the input per se).

The list of receptor systems yielding perplexing small-signal results of this type includes cockroach and spider mechanoreceptors, carotid-body baroreceptors, crayfish and locust stretch receptors, *Limulus* photoreceptors, primate touch receptors, vestibular afferents in birds and mammals, and retinal ganglion cells of the cat. Moreover, in movement-perception systems power-law dynamics are also prevalent; the familiar bell-shaped curve of optomotor response in the beetle *Chlorophanus* has flanks varying, with pattern velocity v, as $v^{0.7}$, as compared with the $v^{2.0}$ of the model with which it has sometimes been compared.

The challenge to the methodologist, therefore, has been to come up with some plausible and testable explanations, given this state of affairs. In the early 1970s, formal procedures were described for mapping power-law dynamics upon sets of more elementary processes having suitable distributions of rate constants. These stemmed from earlier explanations of power-law relaxation in polymers, and of the discharge rates of charged Leyden jars.

The procedure (with examples from photoreception in *Limulus*) is as follows:

1. Formulate the nonlinear differential equations for a plausible candidate transduction process (e.g., for the adapting-bump notions relating light-intensity fluctuation to membrane-conductance change).
2. Linearize the equations for small input changes about a particular operating condition, and examine the resulting linear transfer function, which reveals the physical parameters upon which the small-signal rate constants depend (e.g., transmitter restoration rate, photon capture cross-section, and average local light intensity).
3. Calculate the distributions of values of these physical parameters that would, if a set of such local transduction processes were combined, produce the measured power-law dynamics of the receptor cell (e.g., an approximately exponential spatial distribution of average light intensity along the *Limulus* photoreceptor suffices for the measured small-signal power-law dynamics, and in addition predicts the near-logarithmic steady-state adaptation as average intensity is changed).
4. Use the above inference to test experimentally whether such distributions in fact exist.

This distributed approach has been discussed more recently with regard to photo-, mechano-, and chemoreceptor responses, but clear identification of the distribution of an underlying parameter has—perhaps due to the difficulties inherent in such a measurement—not been reported. Skeptics might even ask why these distributions should just happen to be such that the prevalent power-law dynamics occur. On the other hand, no other sufficient explanation of this ubiquitous dynamic property of sensory receptors has been identified. Some promising work has focused on the separate question of distinguishing between mechanical and electrical origins of the transduction dynamics in mechanoreceptors.

Here are two points of caution for those applying small-signal dynamic analysis to transduction processes: (1) All such systems are of course nonlinear, at least in that the form of the near-linear, power-law behavior can sometimes be altered by changing the steady-state conditions about which it is measured. The fact remains, however, and should not be played down by nonlinear enthusiasts, that a near-linear response under specified conditions requires a near-linear explanation. To our knowledge, no fundamentally nonlinear, nondistributed process has been proposed that, when driven by small perturbations, behaves with the linear fractional-order dynamics measured routinely in sensory transduction. (A separate but related difficulty is that fractional-order differentiation is unfamiliar; it is not hard to find otherwise expert control system engineers who will state mistakenly that the operation is nonlinear. s^k, in the Laplace variable s, is in fact a linear, minimum-phase transfer function.) (2) A rote test for linearity is to see whether responses to step and sinusoidal stimuli are compatible on Fourier-transform grounds. Such tests may fail for reasons that are often neglected, and hence can miss out on some linear inference: Either the step includes Fourier components outside the frequency range of sinusoids tested, or the sinusoidal response is measured only after it has settled down. In the latter case, examination of the sinusoidal response to the first few cycles of the stimulus can sometimes identify a time- or frequency-dependent nonlinearity that can in principle be extracted in the analysis.

More generally, power-law relationships over many decades of the relevant variables abound in nature. These include the decay of phosphorescence, stress relaxation in polymers and muscle, Leyden-jar discharge, size distribution of rubble and dust in the solar system, and measures of the angular spacings of galaxies. My correspondence with workers in each of these fields indicates that they, along with sensory physiologists, face compelling conceptual puzzles like those outlined here.

Further reading

Brown MC, Stein RB (1966): Quantitative studies on the slowly adapting stretch receptor of the crayfish. *Kybernetik* 3:175–185

Chapman KM, Mosinger JL, Duckrow RB (1979): The role of viscoelastic coupling in sensory adaptation in an insect mechanoreceptor. *J Comp Physiol* 131:1–12

Chapman KM, Smith RS (1963): A linear transfer function underlying impulse frequency modulation in a cockroach mechanoreceptor. *Nature* 197:699–700

French AS (1985): Action potential adaptation in the femoral tactile spine of the cockroach, *Periplaneta americana. J Comp Physiol A* 155:803–812

Thorson J, Biederman-Thorson M (1974): Distributed relaxation processes in sensory adaptation. *Science* 183:161–172

Thorson J, White DCS (1983): Role of cross-bridge distortion in the small-signal mechanical dynamics of insect and rabbit striated muscle. *J Physiol* (*Lond*) 343:59–84

Sodium Appetite

Alan N. Epstein

Sodium appetite arises when animals are sodium deficient. It is an increase in their avidity for the taste of salty substances that leads them to seek out such substances and ingest them, and is therefore a specific state of activity within their brains that serves an obvious self-regulatory function. In extreme instances, such as the pathology of adrenal insufficiency or adrenalectomy, the behavior of sodium appetite saves the animal's life.

Sodium deficiency, and therefore the appetite, will occur naturally in animals consuming a diet poor in NaCl, especially in herbivores living on ranges distant from the sea. Female mammals of all kinds, except marine animals, are at risk of the deficiency during pregnancy and lactation, and when they are deficient their milk will be sodium poor and their young may also be deficient. Sodium deficiency may therefore be a common mammalian hazard. It can also be produced by pathological losses or sequestrations of extracellular fluid such as occur with persistent or frequent vomiting or diarrhea, and with congestive heart failure. Experimental treatments (parenteral dialysis with hyperoncotic colloids, caval ligation, constriction of the renal arteries, and parotid fistulation) that reduce plasma sodium or plasma volume induce a sodium appetite, as does pharmacologic paralysis of the renal mechanisms for sodium conservation with natriuretic agents. These treatments have in common a reduction in plasma sodium and consequent increases in circulating angiotensin II and aldosterone which are the hormones of renal sodium conservation.

In the rat, the most commonly studied species, the sodium loss is not the direct cause of the appetite, just as it is not the direct cause of sodium reabsorption by the kidney. Rats that are made sodium deficient do not express an appetite immediately. They do so only hours after the loss has occurred, and they express it at that time despite restoration of plasma sodium to normal levels by withdrawal of it from reservoirs such as bone, or replacement of it by gavage or subcutaneous injection just prior to access to NaCl solutions. The appetite can also be produced by pharmacologic doses of aldosterone or its precursor (DOC, desoxycorticosterone) which, being mineralocorticoids, will maintain the animals in the sodium-replete state. Instead the mechanism of the appetite is, in the rat, an analog of that which causes renal conservation of sodium. That is, it is the hormones of sodium deficiency, angiotensin II and aldosterone, rather than the sodium deficiency itself that apprises the brain of the need for salt and makes it avid for the taste of salty substances. These ideas are schematized in Figure 1 as follows: a sodium deficiency causes the formation of angiotensin II which releases aldosterone, and these two hormones act in synergy on the brain to produce sodium appetite. This is the mechanism for induction of the appetite by natural sodium deficiencies in the endocrinologically intact animal in which both hormones can act. As a result of their synergy they can induce the appetite at concentra-

tions that are within their physiological ranges, and they can do so rapidly. Each of the hormones can induce the appetite by acting alone (as indicated by the parallel arrows from them to the brain), but they do so as the result of pathological states such as adrenalectomy or of pharmacologic treatments such as excess doses of mineralocorticoids.

Experience plays a role in sodium appetite. It is not necessary for its arousal—that is innate. But experience with salty substances prior to sodium deficiency helps the animal find it when it is needed, and makes the expression of the behaviors necessary to gain access to it more rapid. In addition, experience with salt facilitates the recovery from brain damage that either impairs the animal's taste of salt (damage to thalamic gustatory relays) or reduces the animal's urge to drink it (lateral hypothalamic damage). That is, animals that have had an opportunity to find salt in their environment and to sample it do not suffer the impairments in sodium appetite that are produced by these lesions in inexperienced animals.

The central neural apparatus that mediates the appetite is not yet known and may be different from that which permits the need-free animal to express a preference for dilute salt solutions. Judging from lesion and electrical stimulation studies, the central system for the appetite is limbic, involving hypothalamus, septum, and amygdala. Large lateral hypothalamic lesions abolish the appetite, and dorsolateral hypothalamic damage reduces it. Septal damage, which usually produces a hyperdipsia for water, also increases salt drinking. Opposite effects are produced by stimulation of septum and hypothalamus. Amygdaloid lesions can either increase (corticomedial complex) or decrease (medial, basomedial, and basolateral complex) the appetite. Total amygdaloid ablation decreases the appetite but not as totally or as permanently as lateral hypothalamic damage.

Striking species differences are apparent in the mechanisms of the appetite. Sheep are not susceptible to arousal of it by pharmacologic doses of mineralocorticoids. In them, the appetite can be aroused by reduction of brain sodium (by perfusion

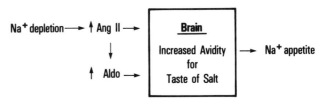

Figure 1. A synergy of angiotensin II and aldosterone are the cause of sodium appetite in the rat.

of the anterior ventricles with mannitol solutions) and can be suppressed by addition of sodium to brain parenchyma by infusion of it into the ventricular space. Sheep appear to be directly sensitive to the sodium content of some crucial tissue close to the third ventricular wall. This is not true of the rat. In addition, wild rabbits become avid for salt when they are treated with a mixture of the hormones of pregnancy (estrogen, prolactin, corticotropin). Further study of the biological basis of sodium appetite is of obvious importance, especially in humans. Such studies could lead to rational therapies for ex-

cesses of salt intake that may predispose to hypertension and that complicate the treatment of renal diseases.

Further reading

DeCaro G, Massi M, Epstein AN (1986): *Thirst and Sodium Appetite*. New York: Plenum Press

Denton D (1982): *The Hunger for Salt*. Berlin: Springer-Verlag

Fitzsimons JT (1979): *The Physiology of Thirst and Sodium Appetite*. Cambridge: Cambridge University Press

Somatosensory Cortex

Jon H. Kaas

Most or all of mammalian parietal cortex is devoted to somatosensory functions. However, the total extent and number of functional subdivisions of this somatosensory cortex varies. Mammals with relatively little neocortex have little somatosensory cortex in addition to the primary and secondary fields. In contrast, monkeys and other higher primates have large amounts of somatosensory cortex, and this cortex contains a number of distinct processing stations.

Somatosensory cortex in primitive mammals

The organization of somatosensory cortex found in tree shrews (Fig. 1), a small, squirrel-like distant relative of primates, is typical of that found in a wide range of mammalian species with relatively little phylogenetic expansion of neocortex. Most of somatosensory cortex in these mammals consists of two distinct systematic representations of the contralateral body

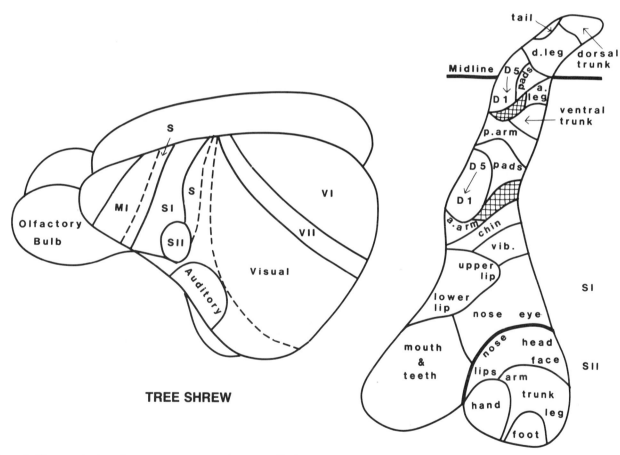

TREE SHREW

Figure 1. The organization of somatosensory cortex in a generalized mammal. Somatosensory cortex in a tree shrew, a distant relative of primates with a relatively small brain, consists of only a few subdivisions. The primary (S-I) and secondary (S-II) areas each systematically represent the body surface as shown. Left. A dorsolateral view of the brain. A narrow strip of cortex (S) along the anterior border and a somewhat wider strip (S) along the posterior border of S-I are also somatosensory, but little else is known about these regions. Other nearby regions of cortex relate to the motor (M-I), auditory, and visual systems. Primary (V-I) and secondary (V-II) visual areas are shown for reference. Right. Details of the body surface representations in S-I and S-II. D1-D5, digits of the hand (lower) and foot (upper). A. arm and leg, anterior arm and leg; p. arm, posterior arm; d. leg, dorsal leg; vib., mystacial vibrissae of the face. Lined regions indicate dorsal surfaces of hand (lower) and foot (upper).

surface, named in order of their discovery in cats, the first (primary) representation, or S-I, and the second representation, or S-II. Almost all neurons throughout both representations respond to gentle stimulation of the skin and body hairs. Thus, S-I and S-II appear to be related largely to cutaneous receptors. Neurons across the thickness of cortex at any location are all activated from the same location on the body surface. However, adjoining groups of neurons across the thickness are activated by different and typically nearby body locations so that all parts of the body eventually are represented in a pattern relative to the cortical surface. The larger S-I represents the body from tail to mouth in a mediolateral cortical sequence, while the smaller S-II has a head-to-tail mediolateral (or dorsoventral) cortical sequence. The two fields join along matched representations of the dorsal midline of the head. Thalamic input to both S-I and S-II is from a single tactile representation, the ventroposterior nucleus. Axon pathways interconnect matched body parts of S-I and S-II of the same cerebral hemisphere and parts of S-I and S-II of the opposite cerebral hemisphere. S-I and S-II also have connections with other regions of cortex, including motor cortex.

Major features of organization of S-I in tree shrews are also common in S-I of other mammals. (1) A large proportion of S-I is devoted to the head. The representation of the head occupies half or more of S-I in tree shrews, squirrels, rats, opossums, rabbits, and several carnivores. Within this head representation, tree shrews and opossums emphasize the glabrous nose, rats emphasize the mystacial vibrissae, and squirrels and rabbits emphasize the lips. Raccoons and monkeys have enlarged representations of the hand. The disproportionate representation of body parts follows the general rule that those skin surfaces with the most important tactile functions have the largest representation (cortical magnification). (2) The representation is largely continuous (somatotopic) so that skin surfaces that are adjacent are usually represented next to each other in cortex. (3) There are often partial disruptions, however, and the locations of these disruptions in the representation vary somewhat across species. For instance, the split representation of the dorsal and ventral trunk is an unusual feature found so far only in tree shrews. In contrast, the representations of the anterior and posterior leg are commonly split in S-I to locations lateral and medial to the representation of the foot. Despite these disruptions, the representations of important sensory skin surfaces are typically continuous. The partial discontinuities seem forced by the requirements of disproportionately representing some skin surfaces, and by the need for representing the three-dimensional body surface within a two-dimensional cortical sheet. (4) Many somatotopic details of S-I in tree shrews are widely found in S-I of other mammals. As in most other mammals, the head is represented with the dorsal surface posterior and the ventral surface anterior in cortex. The hand and foot are represented with the proximal surfaces posterior and the distal digit tips anterior. Digits one-five relate to lateromedial cortical sequences. The ventral trunk is near the posterior border of S-I. In most species, the dorsal trunk is at the same mediolateral level near the anterior border. (5) S-I is associated with cortex characterized by dense packing of small neurons in cortical layer IV (somatic koniocortex). In some mammals (e.g., squirrels and rats), discontinuities in the somatotopic map are associated with discontinuities in the dense packing layer IV cells. As a more marked anatomical specialization in mice and rats, the layer IV cells form ring-like aggregations (barrels) in the part of S-I representing vibrissae, with a barrel for each whisker.

The second somatosensory area, S-II, of tree shrews also has an organization that is typical of that found in other mammals. S-II is orientated so that the top of the head joins the representation of the top of the head in S-I (a congruent border), and the distal limbs point ventrally away from S-I. The somatotopic organization of S-II is cruder than that of S-I in that receptive fields are considerably larger. However, the proportion of S-II devoted to different body parts roughly matches that of S-I. S-II is typically one-tenth to one-half the size of S-I.

Little additional cortex in tree shrews can be considered somatosensory. A narrow fringe of cortex between S-I and motor cortex responds to more intense stimulation of the body, and it is possible that this cortex relates to noncutaneous (deep) body receptors, such as those in muscles, but this is uncertain. A narrow zone of cortex posterior to S-I also responds to more intense stimulation. Cortex posterior to this less responsive somatosensory zone bordering S-I is visual or auditory. Thus, tree shrews appear to have little somatosensory cortex and few subdivisions of somatosensory cortex. Cortical organization is similar in a number of other mammals, including rats, squirrels, and rabbits. However, cats have at least five somatosensory representation, S-I, S-II, an area 3a between S-I and motor cortex that is responsive to muscle spindle receptors, an S-III along the posterior border of S-I, and an S-IV in cortex lateral to S-III.

Prosimian primates

The prosimian primates have less-developed brains than the monkeys and the other advanced primates, apes and humans. Of prosimian primates, the organization of somatosensory cortex has been studied most extensively in galagos. As in tree shrews, much of somatosensory cortex in galagos consists of S-I and S-II. S-I is organized much like S-I in tree shrews, although the representation of the hand is larger. As for tree shrews, S-I in galagos responds to cutaneous stimuli throughout its extent, receives input from the ventroposterior nucleus, and is coextensive with somatic koniocortex. S-I is adjoined lateroventrally by a smaller S-II, which extends into the lateral fissure. As for nonprimate mammals, S-II forms a congruent border with S-I along matched representations of the dorsal midline of the head. The distal limbs in S-II point away from S-I, and the head is in cortex anterior to that devoted to the foot. A narrow strip of cortex on the anterior border of S-I appears to be the homolog of area 3a of monkeys because of its structural similarity and its activation by deep body receptors, possibly muscle spindle receptors. A strip of cortex posterior to S-I responds to more intense stimuli, and this strip may be related to deep body receptors in joints and muscles. Thalamic input to this strip is from neurons dorsal to the ventroposterior nucleus. As in most nonprimates, somatosensory cortex in galagos does not appear to be extensive or to consist of many subdivisions.

Monkeys, apes and humans

Somatosensory cortex in higher primates contains more subdivisions than somatosensory cortex in generalized nonprimates. Most of what is known depends on experiments in monkeys, but it seems reasonable to conclude that somatosensory cortex contains at least as many subdivisions in apes and humans as in monkeys.

Early experiments on the organization of anterior parietal cortex in macaque monkeys defined S-I as a broad region including architectonic areas 3 (3a and 3b), 1, and 2 of Brodmann. The locations of these four architectonic strips are shown for an owl monkey in Figure 2. It is now recognized

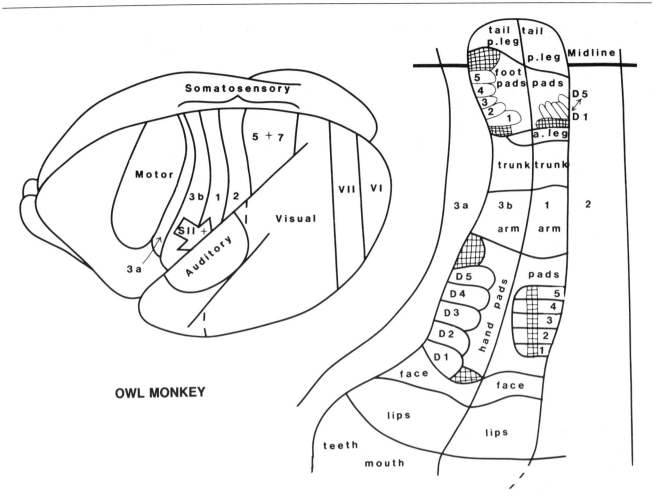

Figure 2. The organization of somatosensory cortex in an owl monkey, a moderately advanced primate. The owl monkey lacks a central fissure so that areas 3a and 3b are exposed on the dorsolateral surface of the brain. Left. A dorsolateral view of the brain. Somatosensory cortex includes the traditional architectonic fields after Brodmann of areas 3a, 3b 1, 2, 5, and 7, and adjoining cortex on the upper bank of the lateral fissure. Cortex in the lateral fissure contains several somatosensory subdivisions of cortex (S-II+) including S-II. The locations of visual cortex, including primary (V-I) and secondary (V-II) fields, auditory cortex, and primary motor cortex are shown for reference. Right. Details of the body surface representation in areas 3b and 1. D1-D5, digits of the hand (lower) and foot (upper). a. and p. leg, anterior and posterior leg.

from detailed microelectrode experiments on owl and other monkeys that each of the four architectonic fields—3a, 3b, 1, and 2—contains a separate representation of the body. Since these representations roughly match each other in mediolateral cortical sequences, early experiments using grosser recording techniques failed to distinguish them.

Because of the historical definition of S-I in monkeys, it is common to include areas 3a, 3b, 1, and 2 in S-I. However, only area 3b appears to be the homolog of S-I as generally defined in nonprimates and prosimians. Like S-I, area 3b is coextensive with cortex distinguished by densely packed small cells in layer IV, has dense input from the ventroposterior nucleus of the thalamus, is responsive throughout to light tactile stimuli, and has digit tips of the hand and foot represented along the rostral border. Area 3b forms a complete representation of the body surface. Major cortical projections of area 3b are to S-II and area 1.

Area 1 also contains a complete representation of the body surface. The somatotopic organization of area 1 is roughly a mirror reversal of that in area 3b (Fig. 2). Thus, the digit tips of the hand and foot are represented posteriorly in area 1, and the somatotopic pattern is the reverse of that found in

S-I of galagos and nonprimates. In comparison to area 3b, layer IV is less densely packed with small cells and the input from the ventroposterior nucleus is less dense. Area 1 also differs from area 3b by having fewer neurons with slowly adapting responses to maintained stimuli, neurons with larger and more complex receptive fields, and some neurons that appear to be activated by the rapidly adapting Pacinian receptor class. While area 1 is architectonically apparent in the cortex of monkeys, apes, and humans, there is no compelling evidence for this field in prosimians and nonprimates. Thus, it appears that the area 1 representation exists, or at least it is well developed, only in higher primates.

Area 3a is characterized by less dense packing of small layer IV cells and more layer V pyramidal cells than area 3b. Area 3a is activated predominantly by muscle spindle receptors and receives input from an anterodorsal division of the ventroposterior thalamic complex. The details of the somatotopic organization of area 3a have not been fully determined, but the field represents deep receptors in parallel with the adjoining representation of cutaneous receptors in area 3b. Similarities in responsiveness, histological structure, and even connections of area 3a in monkeys to the field termed

3a in cats are striking. Since a comparable 3a field has not been identified in a range of mammalian species, many of these similarities may be the result of parallel evolution. However, the reported presence of a somatosensory fringe along the anterior border of S-I of a number of mammals suggests that a less developed 3a field exists in many or most mammals.

Area 2 occupies a strip of cortex immediately caudal to area 1. Layers IV and VI are more densely packed with cells than area 1. The posterior boundary of area 2 is less apparent, and there is disagreement on the location of this border. The somatotopic organization of area 2 is incompletely understood, but much or all of area 2 contains a body representation that is parallel to that in area 1, and at least partly mirrors that in area 1. Area 2 is activated by both cutaneous receptors and receptors in deep body tissues (muscles, joints, and tendons). The information from deep receptors appears to be relayed from dorsal divisions of the ventroposterior thalamic complex, and perhaps the anterior pulvinar, while information from cutaneous receptors appears to be relayed from areas 3b and 1 and, to a lesser extent, from the ventroposterior nucleus. Many neurons in area 2 are activated by both skin and deep receptors, and many respond best to complex stimuli involving shapes or direction of movement. Thus, area 2 seems to be a somewhat higher order processing station. There is no clear evidence for a homolog of area 2 in prosimians or nonprimates. However, in a number of mammals, cortex posterior to S-I receives input from part of the thalamus that is dorsal to the ventroposterior nucleus, and input from this location is suggestive of the input to area 2 from the dorsal parts of ventroposterior complex in monkeys.

Somatosensory cortex in monkeys also includes a large region of cortex on the upper bank and insular surface of the lateral fissure and the large posterior parietal region of cortex. The lateral fissure contains several somatosensory areas including an S-II proper that appears to be homologous to S-II in nonprimates. As in other mammals, S-II in monkeys responds to cutaneous stimuli and forms a systematic representation of the contralateral body surface. Cortical inputs are from areas 3b, 1, and 2 and thalamic inputs appear to be largely from the ventroposterior inferior nucleus (VPI) and perhaps somewhat less from the ventroposterior nucleus (VP). Pacinian receptors may relate to part or all of S-II (or a subdivision of cortex outside of S-II proper) via VPI. One recent concept of S-II function in monkeys, and thereby in other mammals, is that S-II provides a hierarchical link in a relay of somatosensory information to limbic cortex that is critical in the permanent learning of tactile discriminations.

The organization of cortex adjoining S-II in monkeys is not completely understood. However, cortex outside of S-II in the lateral fissure responds to somatosensory stimuli and contains subdivisions with different architectonic characteristics, types of neurons, and connection patterns with other fields and thalamic nuclei.

Posterior parietal cortex has been subdivided in various ways, but most investigators refer to area 5 and area 7 of Brodmann. These subdivisions are not completely satisfactory because there is no general agreement as to the exact locations and boundaries of cytoarchitectonic fields or on the criteria that defines them. In addition, and very importantly, neither field seems to be functionally homogeneous. Thus, the concepts of areas 5 and 7 have limited usefulness, and ultimately they may be redefined or replaced by more precisely defined subdivisions. Nevertheless, some general conclusions are possible. The more anterior part of posterior parietal cortex (area 5) receives cortical inputs from areas 1 and 2, and area 5 relays to more posterior parietal cortex (area 7). Area 7 also receives inputs from the lateral posterior nucleus of the thalamus and subdivisions of extrastriate visual cortex. The area 7 region appears to have higher order functions since the response properties of neurons are complex and include types related to vision, eye movements, and visually guided movements. Lesions of area 7 produce visuospatial disorientation, defects in eye movements, misreaching, and a reluctance to use contralateral limbs.

Modular organization

In early landmark studies of somatosensory cortex, Mountcastle proposed that cortical areas are not uniform in function, but are divided into small, functionally distinct processing modules, each containing a limited group of cells in a column extending from the surface of the cortex to fibers below. This concept has been widely accepted, but there have been only limited efforts to reveal the nature of processing units in any region of somatosensory cortex. The general validity of the concept is strongly supported by anatomical studies of connections, especially corticocortical connections, that cells in one location in the somatosensory system often project to a number of closely spaced separate foci in another area, thus suggesting a spacing of neurons involved in a particular function.

Recent electrophysiological evidence for one type of modular organization came from experiments where the local distributions of slowly adapting and rapidly adapting cortical neurons were determined within area 3b of monkeys by recording from successive neurons across the depth of cortex in many closely spaced microelectrode penetrations. In the representation of the glabrous hand in area 3b of monkeys, most or all neurons respond to slight indentation of the skin with a probe. Some neurons respond only to the onset and offset of the indentation (rapidly adapting) while others continue to respond during a maintained indentation (slowly adapting). The slowly adapting neurons appear to be largely confined to the middle layers of cortex where they are clustered in extended foci that appear to form closely spaced bands coursing parallel to the surface. Cells in superficial and deep layers over and under these slowly adapting bands are rapidly adapting, and cells between the slowly adapting bands in layer IV form rapidly adapting bands. The reasons for this distribution are not clear. One possibility is that inputs relayed from slowly adapting and rapidly adapting receptor classes in the skin distribute in alternating bands in layer IV of cortex, and that slowly adapting responses are inhibited in cells outside the receiving layers. However, the observations demonstrate that processing modules in somatosensory cortex may be more band-like than column-like in form, and may be related to specific cortical layers, as are orientation-selective bands and ocular dominance bands in visual cortex.

The reorganization of cortical maps after injury

The somatotopic organization of areas of somatosensory cortex do not reflect the full potential of the anatomical connections, and local features of organization in these maps can be altered by small lesions in the representations or by removal of the source of an input to part of a representation. Thus, cutting a sensory nerve to part of the hand or removing a finger does not result in a permanent silence of the cortex corresponding to the former representation of part of the hand or a finger. Rather, over a short period of hours to weeks, neurons throughout the deprived cortex become responsive to stimuli on the normally innervated parts of the hand. Likewise, following a small lesion of cortex that removes the representation of a given finger, neurons in cortex representing adjacent fingers come to be activated by inputs from the finger with the removed representation. These experiments indicate that somatotopic representations in adult mammals are subject to small degrees of change and reorganization, and that many local features of representations are dynamically maintained. Because changes can be quite rapid, the anatomical connections mediating the reorganization appear to be at least roughly in place, and the reorganization process appears to be one of changing the effectiveness of previously existing anatomical pathways.

Further reading

Kaas JH (1983): What, if anything, is SI? Organization of first somatosensory area of cortex. *Physiol Rev* 63:206–231

Kaas JH, Merzenich MM, Killackey HP (1983): The reorganization of somatosensory cortex following peripheral nerve damage in adult and developing mammals. *Ann Rev Neurosci* 6:325–356

Kaas JH, Nelson RJ, Sur M, Merzenich MM (1981): Organization of somatosensory cortex in primates. In: *The Organization of the Cerebral Cortex*, Schmitt FO, Worden FG, Adelman G, Dennis SG, eds. Cambridge: MIT Press

Hyvarinen J (1982): Posterior parietal lobe of the primate brain. *Physiol Rev* 62:1060–1129

Robinson CV, Burton H (1980): Organization of somatosensory receptive fields in cortical areas 7b, retroinsula, postauditory and granular insula of *M. fascicularis*. *J Comp Neurol* 192:69–92

Sound Communication in Anurans (Frogs and Toads), Neuroethology of

Robert R. Capranica

There are almost 3,000 species of frogs and toads in the world today. Even though a number of different species may share a common breeding site, hybridization is relatively rare. The basis for this reproductive isolation resides in the males' species-specific mating (advertisement) calls. Females are attracted selectively to calling males of their own species. These observations indicate that anurans are capable of precise acoustic recognition of complex sounds in a noisy environment, since the sound level of each calling male in a mixed chorus often is of the order of 100–110 decibels SPL (sound pressure level) at a distance of 1 meter. Furthermore, despite the small interaural distance between their eardrums and lack of external pinnae, they show remarkable accuracy in sound localization (which has been quantified by observation of female phonotactic approaches to playback of the male's call.) Because of the fundamental role that vocal signals play in their lives and the fact that auditory discrimination and sound localization can be studied by means of acoustic playback of natural and synthetic calls (electronically or computer generated), anurans have proved to be a model system for neuroethological studies of sound communication.

Males of most species possess a simple, subgular inflatable vocal sac beneath their throats; some species, particularly in the genus *Rana*, may possess paired lateral vocal sacs. The signal characteristics of the mating call depend primarily on laryngeal mechanisms and not on the vocal sac itself. That is, the vocal sac serves primarily as an acoustic radiator that enables the efficient broadcast of acoustic energy into the environment. Vibration of the vocal cords converts the steady pulmonary pressure into alternating pressure cycles. Tension on the vocal cords is under neural control, thus enabling rapid changes in frequency modulation of vocal cord vibration rates in some species. Amplitude modulations can be superimposed by separate vibration of the arytenoid cartilages that overlie the vocal cords. Anurans therefore possess a larynx that can give rise to a wide diversity of species-specific sounds.

Most anurans possess a pair of circular tympanic membranes (eardrums) symmetrically located on each side of the head. Vibrations of an eardrum are coupled to the inner ear by means of columellar middle-ear apparatus. Within the fluid-filled inner ear are two separate and distinct acoustic receptor organs: amphibian papilla and basilar papilla. Amphibians are therefore unique in this regard; all other vertebrate classes possess a single auditory organ. The amphibian papilla is tuned to lower tonal frequencies than the basilar papilla. The exact tuning of the two organs varies among species and matches the spectral energy distribution in sounds of biological significance, particularly the mating call. Thus the sensitivity of the inner ear of anurans is specialized for selective reception of particular sounds. The presence of peripheral specialization and its role in processing complex sounds is not so obvious in other vertebrate classes, such as birds and mammals.

Despite the lack of a movable basilar membrane in either organ, auditory nerve fibers possess typical "tuning curves" with individual best excitatory frequencies and other response properties that are remarkably similar to those in more advanced vertebrates. Auditory nerve fibers terminate in a dorsal nucleus in the medulla which in turn projects to the superior olivary nucleus and thence to the torus semicircularis in the midbrain. The ascending auditory nervous system therefore resembles the "wiring diagram" of other tetrapods. Neurons in these ascending centers exhibit an increasing complexity in their response patterns to frequency and temporal features, as well as binaural sensitivity to interaural time and intensity differences, which reflects a hierarchical scheme in processing complex sounds. Many cells in the forebrain (thalamus and telencephalon) are selective for combinations of spectral and temporal features contained in the calls within a species' vocal repertoire. These results indicate that higher auditory centers play an important role in the detection and recognition of species-specific acoustic communication signals.

Further reading

Capranica RR (1976): Morphology and physiology of the auditory system. In *Frog Neurobiology,* Llinas R, Precht W, eds. Berlin: Springer-Verlag, pp 551–575

Wilczynski W, Capranica RR (1984): The auditory system of anuran amphibians. *Prog Neurobiol* 22:1–38

Sound Localization in the Owl

Masakazu Konishi

Finding the source of a sound by ear is called sound localization, and most animals, including nocturnal owls, that use sound signals for communication, prey capture, and predator evasion can localize sound. Sound localization and its brain mechanisms are better known in the barn owl than in any other animal. The barn owl can localize sound equally well in two dimensions, azimuth, or the horizontal plane, and elevation, or the vertical plane. Experiments show that the barn owl uses interaural time difference for determining sound azimuth and interaural intensity difference for elevation. When a sound signal reaches one ear before the other ear, an interaural time difference results. Sound travels 1 cm in 29 μsec. The maximum interaural time difference experienced by the barn owl is about 170 μsec when sound propagates along the aural axis; the minimum time difference is zero when sound emanates from a place equidistant to the two ears. Thus, interaural time disparity varies systematically as a function of sound azimuth. Interaural intensity difference is due to the head's shadow in the sound field. The ear in the shadow registers a weaker signal than the other ear. Given a constant head size, shorter wavelengths and thus higher frequencies produce greater intensity differences than longer wavelengths. The barn owl needs relatively high frequencies (5–8 kHz) for accurate localization in elevation. Because the left ear is situated higher on the head than the right one in the barn owl, interaural intensity disparity varies as a function of sound elevation.

The barn owl derives both interaural cues from the same high frequency signal. However, its auditory system processes the two cues in two separate pathways within the brain stem. The owl's ear performs spectral analysis of sound and encodes the phase and amplitude of each frequency component.

Birds have two anatomically separate cochlear nuclei on each side, nucleus angularis and nucleus magnocellularis, and each primary auditory fiber divides into two collaterals, one innervating nucleus angularis and the other nucleus magnocellularis. Although the phase and intensity codes are carried by both collaterals, nucleus angularis accepts only the intensity code and nucleus magnocellularis only the phase code. Neurons of nucleus angularis are sensitive to variation in sound intensity but insensitive to stimulus phase, whereas magnocellular neurons phase-lock to tonal stimuli but are insensitive to variation in sound intensity. Thus each cochlear nucleus is specialized to process one cue but not the other.

The two pathways that start from the cochlear nuclei are anatomically separate until they converge in the inferior colliculus. The two channels are referred to as the time and intensity pathways. The time pathway contains designs necessary for the resolution of interaural time disparities in the microsecond range in high-frequency signals. Magnocellular neurons phase-lock to tonal stimuli of high frequencies (5–8 kHz), which are much too high for mammalian auditory neurons to phase-lock. Nucleus magnocellularis projects bilaterally to a third-order station, nucleus laminaris, which is presumably the avian homolog of the superior medial olivary nucleus. Thus, nucleus laminaris is the first site of binaural convergence in the time pathway. The neural circuits containing magnocellular and laminaris neurons form delay lines; the phase codes from the two magnocellular nuclei reach laminaris neurons with different delays even when the sound signal stimulates the two ears simultaneously. Furthermore, laminaris neurons are coincidence detectors; they fire maximally when the phase codes arrive simultaneously from the left and right magnocellular nuclei. Therefore, when a lead in acoustic transmission cancels a neural delay, the neurons having that delay fire maximally. Thus, delay lines and coincidence detection by these neural circuits are used to measure interaural time difference. Because a unique delay is assigned to a neuron, each neuron is tuned to a specific range of interaural disparities. Different laminaris neurons show different delays, ranging from 0 to about 170 μsec. Moreover, laminaris neurons are systematically arranged to form a map of delays in each frequency band. The repetition of the map in each frequency band is due to the time and phase variation with frequency or period. The owl's auditory nerve and nucleus magnocellularis preserve the phase of all frequencies in a signal for binaural comparison in nucleus laminaris. The delay lines in different frequency channels are adjusted so as to derive a single interaural time difference from many interaural phase differences.

Nucleus laminaris projects both to the inferior colliculus directly and to one of the lemniscal nuclei. Neurons of the lemniscal nucleus are also tuned to a specific range of interaural time differences. The time pathway ultimately reaches the external nucleus of the inferior colliculus. Neurons of the external nucleus are much more sharply tuned to a specific range of interaural time disparities than those of either the lemniscal or laminaris nucleus, indicating the presence of processes for sharpening the selectivity of neurons for time disparities. Another mechanism that occurs in the ascending time pathway is the elimination of ambiguities in the derivation of a time difference from a phase difference. Because the phase difference preferred by a neuron recurs in every tonal cycle, the neuron responds to it in every cycle; the neuron cannot determine which of the phase differences corresponds to the correct time difference. Although the barn owl shows phase ambiguity with tonal stimuli, it does not with noise stimuli. In the time pathway all neurons below the external nucleus of the inferior colliculus show phase ambiguity even to noise stimuli. However, neurons of the external nucleus do not show phase ambiguity to noise signals.

Little is known about the processing of interaural intensity difference in the intensity pathway. The time and intensity pathways converge on the neurons of the external nucleus.

These neurons respond only to a specific combination of inter-aural time and intensity disparities, either cue alone being ineffective. Because such a combination occurs only in a partic-ular area of the owl's auditory space, the neurons tuned to the combination respond only to sound emanating from the area, which is referred to as a receptive field. Furthermore, these space-specific neurons are systematically arranged ac-cording to the location of their receptive fields, thus forming a neural map of auditory space. Interestingly, the map projects topographically onto the optic tectum where visual-auditory neurons form a bimodal map of the owl's bisensory space.

Further reading

Konishi M, Sullivan WE, Takahashi T (1985): The owl's cochlear nuclei process different sound localization cues. *J Acoust Soc Am* 78:360–364

Sullivan WE, Konishi M (1984): Segregation of stimulus phase and intensity in the cochlear nuclei of the barn owl. *J Neurosci* 4:1787–1799

Takahashi T, Moiseff A, Konishi M (1984): Time and intensity cues are processed independently in the auditory system of the owl. *J Neurosci* 4:1781–1786

Spinal Cord, the Dorsal Horn

Richard E. Coggeshall

The mammalian spinal cord is made of white matter and gray matter. The white matter consists predominantly of myelinated axons, whereas the gray matter is a neuropil, a complex tangle of cells, axons, dendrites, synapses, supporting cells, and blood vessels. On cross-section, the gray matter appears as a thickened and distorted letter H, and the upper arms or tines of the H are referred to as the dorsal horns. This term is a misnomer, however, for it is only on cross-section that these structures appear like horns. In reality, since the spinal cord is a long cylinder, the dorsal gray matter forms two shelves that extend the length of the spinal cord.

The dorsal part of the spinal cord is an area for sensory integration. The dorsal root attaches to the spinal cord near the apex of the dorsal horn, and this root contains the segmental sensory axons. A distinction is usually made between large and small sensory axons. The larger fibers enter the dorsal white column and the smaller enter the tract of Lissauer. These fibers or their collaterals enter the gray matter of the spinal cord, and the majority synapse in the dorsal horn. An understanding of the modifications of somatic sensory information in the dorsal horn is one of the major goals of modern spinal sensory research.

The dorsal horn has been divided in various ways. The outer layer (lamina) of the dorsal horn consists of prominment large neurons. This is known as the layer of Waldeyer. Deep to this is a region that contains few myelinated axons and glial processes. For this reason it has a gelatinous appearance and is referred to as the substantia gelatinosa. There was little agreement as to the deeper part of the dorsal horn, however, until Rexed pointed out that the entire gray matter of the spinal cord consisted of layers or laminae. The laminae are recognized by characteristic arrangements of the neuronal cell bodies. The dorsal horn of the spinal cord consists of laminas I–VI.

Unfortunately, few functional generalizations can be made about this area because it is so complex that we have only elementary insights as to how information, particularly sensory information, is being organized.

When an area of the nervous system is not well understood, it is useful to describe the components in terms of inputs, integrating cells, and outputs. There are two major inputs into the dorsal horn, the primary afferent fibers and the descending fibers (from the brain stem and above). There are two types of neurons that react to this input, the propriospinal neurons and the long tract cells. The propriospinal neurons are neurons whose cell bodies and processes are completely confined within the spinal cord. The long tract cells are neurons whose cell bodies are located in the spinal cord but whose axons extend to the brain stem or brain. The output of the dorsal horn is to the brain stem and brain via the axons of the long tract cells and to motor areas of the spinal cord via the axons of propriospinal cells. The propriospinal neurons and their processes obviously occupy a key position in the transfer of information in the dorsal horn, and an important task for students of the spinal cord is to add more precision to our understanding of the function of propriospinal neurons.

One of the characteristics of the organization of the dorsal horn is that it is organized into layers (laminas). An important question is whether these laminas have meaning beyond the relatively obvious function of allowing investigators to locate areas of the spinal cord with precision. Recent evidence indicates that different types of afferent input are received by different laminas within the dorsal horn. Thus, the laminas are meaningful in terms of different types of sensory input, as well as being important landmarks.

Probably the most intense clinical interest in the dorsal horn concerns the problem of pain. One of the disabling features of spinal cord damage is disordered and sometimes intensely magnified pain. One way to deal with this problem, if the pain is segmental, is to remove all sensory input from that part of the body that is suffering. This will remove the pain at the cost of complete anesthesia, but this is not an ideal solution. Other operations and procedures also lessen the pain without affecting other types of sensation, but all these procedures have flaws. It seems fair to say that as our understanding of the basic organization of the dorsal horn of the spinal cord progresses, more useful therapies for spinal damage will result.

Further reading

Willis WD, Coggeshall RE (1978): *Sensory Mechanisms of the Spinal Cord.* New York and London: Plenum Press

Taste

Lloyd M. Beidler

Many living forms, from bacteria to humans, can detect and discriminate a large number of chemicals in their environment. The major function is associated with food selection and consumption. In many species of fish, amphibians, humans, and other mammals, the sense organs of taste are clusters of cells called taste buds. In mammals they are found primarily on the tongue and soft palate with a smaller number in other areas of the oral cavity.

A single taste nerve fiber innervates several cells within a taste bud as well as those of close neighbors. Transection of a taste nerve bundle results in taste bud degeneration and ultimate elimination. Nerve regeneration is followed by reappearance of taste buds.

A single taste cell alters its resting potential in response to adequate chemical stimulation. A variety of chemicals associated with the taste qualities of sour, salty, bitter, and sweet may stimulate a single taste cell, although no two cells present identical responses to a series of such chemicals. Likewise, a single taste fiber may respond to stimuli associated with more than one of these taste qualities, although many fibers may respond best to but one of them.

Taste buds are not uniformly distributed over the tongue, and certain areas are more sensitive to one taste quality than another. In humans the taste buds of the circumvallate papilla at the back of the tongue are sensitive to bitter substances. Sweet stimuli are often best perceived on the tip of the tongue, sour at the sides, and salty on the edges.

The relationship between the structure of the chemical stimulus and the taste quality it elicits is quite complex. Sour taste is primarily associated with hydronium ions, although the anion may determine the magnitude of sourness. Most salts have mixed tastes, with bitterness being predominant with those of large molecular weights. Chloride salts of sodium and lithium elicit a predominantly salty taste. Lead acetate has a sweet taste. The taste quality of all salts varies with concentration. Many chemicals can elicit a sweet taste. These include sugars, sugar derivatives, glycols, alcohols, some amino acids and dipeptide esters, as well as dulcin, cyclamate, and saccharin. A necessary but not sufficient requirement for a sweet compound is an AH-B moiety. The distance between the hydrogen atom and the negative B atom is about 3 A. In addition to hydrogen bonding to the receptor at these two sites, a third hydrophobic site is utilized for dispersion bonding. These three sites determine the stereospecificity associated with many sweet compounds.

Diverse groups of substances such as lactoses, oximes, peptides, urea and thiourea derivatives, glycosides, and thio carbonyls exhibit bitterness. In contrast to sweet substances, bitters appear to have many different structural groups such as polar moieties responsible for their tastes. Lipophilic properties are also important for a number of bitter chemicals. It is interesting to note that small changes in structure of sweet molecules often alter their tastes and they become bitter.

After taste cells are stimulated in the oral cavity, the first-order nerves are excited and the neural messages are transmitted to the nucleus of the solitary tract of the medulla. Second-order gustatory neurons project to the parabrachial nuclei of the pons. In the rat, third-order neurons project to the gustatory cortex and the stria terminalis, amygdala, and hypothalamus of the forebrain. Efferents from the gustatory and visceral systems of the forebrain project back to the gustatory areas of the pons and medulla. Thus, these forebrain sensory systems may be involved in ingestive behavior and regulation of the animal's energy and fluid intake.

Many natural carbohydrates are sweet and toxic substances bitter. Since most animals accept sweet and reject bitter foods, their tastes appear to play an important role in the monitoring of food intake. Radiation or chemically induced sickness, experimentally produced and paired with ingestion of a well-defined tastant, often results in a long-lasting aversion to the specific tastant. In addition to their role in taste perception, taste buds may also initiate reflex actions that modulate internal levels of chemicals involved in the body's nutrition and metabolism.

Further reading

Beets MGJ (1978): *Structure-activity Relationships in Human Chemoreception*. London: Applied Science Publishers

Beidler LM, ed (1971): *Handbook of Sensory Physiology. Vol 4. Chemical Senses. 2. Taste*. Berlin: Springer-Verlag

Pfaff DW, ed (1985): *Taste, Olfaction, and the Central Nervous System*. New York: Rockefeller University Press

Taste, Psychophysics

Linda M. Bartoshuk

Taste psychophysics is the branch of experimental psychology that deals with the quantitative analysis of taste experience. Fechner's (1801–1887) classic methods permitted the measurement of the lowest concentrations just detected as having a taste (detection thresholds) or recognized to have a specific taste quality (recognition thresholds). Modern psychophysical investigations still use these classic methods (modified to incorporate the insights of signal detection theory), but scaling of perceived intensity has become an increasingly popular psychophysical tool.

Theoretical debates about the validity of various scaling methods abound; however, the direct methods introduced by S.S. Stevens are used widely in taste. Taste conforms relatively well to Stevens's power law. That is, magnitude estimates of perceived taste intensity are roughly equal to the concentration of the substance raised to a given power. This latter value, the exponent of the power function, varies with the nature of the stimulus, its temperature, and the method by which it is tasted. In taste, exponents vary from less than to greater than 1.0. For example, sucrose, at room temperature tasted as one normally sips a beverage, produces a psychophysical function with an exponent of 1.0 or greater. This means that doubling the concentration will double the perceived sweetness (exponent of 1.0) or may even more than double the perceived sweetness (exponent > 1.0). However, if the sucrose is warmed to body temperature and flowed gently across the extended tongue, the exponent will be considerably less than 1.0. In this case, doubling the concentration will produce a sweetness that is less than double that of the original concentration.

There is still controversy over the nature of taste qualities. Most investigators accept the four basic tastes: salty, sweet, sour, and bitter. Some argue that these four are not exhaustive and still others argue that the four tastes are really only points on a continuum of qualities. In spite of the theoretical arguments, most psychophysical studies use stimuli that are exemplars of the four tastes.

The independence of the four taste qualities led Öhrwall, one of the great taste psychophysicists of the turn of the century, to propose that they should be considered as four separate modalities. Although never popular, possibly because of the opposition of Kiesow, another great taste investigator, and a student of Wundt, modern taste studies support Öhrwall's ideas. The most important of these concern adaptation, the fading of taste sensation with constant stimulation. Under the proper conditions (moderate concentrations, tongue held still), adaptation can be complete in taste. That is, taste sensation can completely disappear. To reinvoke the normal taste of the substance, the concentration must be increased to a value higher than the adapting concentration. All concentrations lower than the adapting concentration will take on an untypical quality, called a ''water taste.'' For example, following adaptation to NaCl, citric acid, sucrose, quinine, and urea, water tastes bitter-sour, sweet, bitter, sweet, and salty, respectively. The artichoke provides an especially dramatic example for some individuals. After eating globe artichokes, water tastes sweet for up to 15 minutes.

Cross-adaptation refers to the ability of adaptation to one substance to reduce the taste intensity of a different substance. If cross-adaptation occurs between two substances, they are believed to have common receptor mechanisms. Cross-adaptation does not occur across substances with different taste qualities. Cross-adaptation is complete (for the substances tested so far) for the salty taste and for the sour taste. For example, adaptation to one acid renders them all less sour. Cross-adaptation is not complete across all bitter and all sweet substances. For example, adaptation to aspartame partially cross-adapts sucrose but not saccharin.

The independence of the taste qualities can be demonstrated with the aid of some rather exotic plants. Topical application of a tea made from *Gymnema sylvestre* leaves can abolish sweetness without abolishing the other qualities. Topical application of *Synsepalum dulcificum* berries induces an intense sweet taste in the presence of acid. Some less exotic modifiers, anesthetics and detergents, also have selective effects on taste qualities.

The most impressive interaction among different qualities occurs in mixtures. When two substances with different taste qualities are mixed, both qualities are usually suppressed. However, the qualities remain recognizable unless one is suppressed so severely that it disappears. Some of this suppression appears to result from the ability of one substance to interfere with the initial binding of another. In other cases, suppression appears to occur in the central nervous system.

Genetic differences exist in taste ability. Thresholds for PTC (phenylthiocarbamide) are bimodally distributed. Family studies suggest that the inability to taste PTC is inherited as a simple Mendelian recessive. Those with low thresholds carry either one tasting gene (heterozygous) or two tasting genes (homozygous) while those with high thresholds carry two nontasting genes. The taste worlds of tasters and nontasters of PTC differ with regard to more than the PTC compounds themselves. A variety of bitter and sweet compounds taste more intense to tasters.

Further reading

Bartoshuk LM (1978): Gustatory system. In: *Sensory Integration, Vol 1, Handbook of Behavioral Neurobiology.* New York: Plenum Press

McBurney DH (1974): Are there primary tastes for man? *Chem Senses* 1:17–28

McBurney DH (1978): Psychological dimensions and perceptual analyses of taste. In: *Handbook of Perception,* Carterette EC, Friedman MP, eds. New York: Academic Press

Taste and Smell Disorders

Robert I. Henkin

Incidence

Taste and smell dysfunction reflect a panoply of symptoms affecting many people with varying intensity and diversity. Following a recent U.S. survey of smell, it was estimated that 16 million people exhibit some form of taste or smell dysfunction. Approximately 70% of patients exhibit four major diagnostic categories of disease; 25% have postinfluenza-like hyposmia and hypogeusia (PIHH), an illness which usually follows a severe coryza; 20% have an idiopathic cause (10% of these patients have some form of malignancy, commonly occult); 15% have had a head injury; and 12% have allergic rhinitis most commonly altering smell function.

Symptoms

Hyposmia (decreased smell function) is four times as common as hypogeusia (decreased taste function), although patients commonly complain they cannot taste when, in reality, they mean that they are not obtaining adequate flavor from food, reflecting decreased smell, not taste. Ageusia, the total loss or absence of taste function (i.e., the inability to perceive salty, sweet, sour, acid, or bitter tastants) is rare, as is anosmia, which occurs in about 2% of patients with smell dysfunction. Approximately 50% of patients with hypogeusia or hyposmia also exhibit dysgeusia, a distortion of taste. Cacogeusia describes the obnoxious taste associated with the intake of food and beverages usually considered pleasant; phantogeusia, a taste present in the oral cavity in the absence of any food or drink; heterogeusia, the similar, distorted taste of all food and drink (e.g., all food and drink may taste rotten); and parageusia, an unusual but not obnoxious taste distortion, e.g., sugar tastes like salt. Dysosmia describes similar disorders of smell consisting of cacosmia, the obnoxious smell of vapors usually considered pleasant, phantosmia, the presence of a smell in the nasal cavity without the presence of an external stimulus (commonly a burnt or chemical smell), heterosmia, or parosmia. The various forms of dysgeusia and dysosmia may also occur alone without hypogeusia or hyposmia. Dysgeusia has been categorized into four types (I–IV) based upon the number of foods and beverages perceived as distorted—the greater the number of distorted foods and beverages, the greater the weight loss, nutritional deficits, and nutritional risks.

Pathology

Both taste and smell systems involve three major components: receptors, where the sensory signals are received and the chemical energy transduced into electrical signals; nerves, over which the sensory signals are transmitted; and the brain, where the sensory signals are integrated. Abnormalities of any of these components can result in malfunction of the system from either direct or indirect insults. For example, direct impairment of smell may result from a tumor of the olfactory bulbs (neuroesthesioblastoma or olfactory groove meningioma); of taste, from a leukemic infitrate of the lingual surface. An indirect impairment may involve metabolic abnormalities that alter body chemistry and affect several organ systems including those of taste and smell (e.g., endocrine, trace metal, or nutritional abnormalities). Drug reactions are common initiants of taste and smell dysfunction. Local pathology in the nasal cavity (e.g., nasal polyps) can impair vapor delivery to the olfactory epithelium; local pathology of the oral cavity (e.g., xerostomia secondary to salivary gland dysfunction) can result in taste bud destruction and subsequent hypogeusia.

Diagnosis

Hypogeusia has been defined quantitatively by measurement of taste function. Clinically, this has been defined by measurement of detection and recognition thresholds, most commonly by use of the forced-choice, three-stimuli drop technique and by measurements of magnitude estimation that define the magnitude and character of the hypogeusia. The bitter taste quality is most commonly affected by pathological changes, the sweet quality the least. Hyposmia is defined by measurements of smell function with measurement of thresholds (by the standard forced-choice, three-stimuli sniff technique) and with magnitude estimation for various vapors. Detection thresholds are most commonly affected by pathological conditions, particularly for vapors usually appreciated as pleasant. Impaired magnitude estimation implies loss of receptor number for smell as well as for taste. Smell dysfunction is categorized as type I hyposmia (a qualitative inability to recognize vapors at the primary olfactory area, involving the interaction between the olfactory epithelium and the olfactory nerves), type II hyposmia (a quantitative decrease in ability to detect or recognize vapors), or anosmia (the inability to detect or recognize vapors at either primary or accessory olfactory areas).

To diagnose these complex problems, evaluation of pathological changes in several organ systems are necessary. A complete medical history and physical examination is important, with special attention paid to the neurological system and to the head and neck areas. Evaluation of hemotological, renal, liver, endocrine, and other metabolic systems is important, at least by screening blood studies. Allergic manifestations may be indicated by use of serum IgE or RAST tests. Roentgenograms of the chest and sinuses are useful. Computerized tomography scans and magnetic resonance imaging studies of the brain can reveal specific central nervous system pathology, and positron emission tomography scanning is useful to identify specific functional lesions in the brain.

Treatment

Treatment, as in other conditions, depends upon establishment of an appropriate diagnosis. If indirectly affected, treatment of the underlying disease usually corrects the taste and smell abnormalities. For example, treatment of hypothyroidism with replacement hormone usually corrects the hypogeusia, hyposmia, dysgeusia, and dysosmia, as it does the other symptoms of the disease, the mechanism in taste and smell related, in part, to thyroid hormone effects on cyclic adenosine monophosphate (cAMP) phosphodiesterase activity and impaired receptor transduction of both gustatory and olfactory signals. For direct effects relief of local pathology (e.g., nasal polypectomy, treatment of lingual or palatal tumors) may restore function, but responses can be complicated by associated disease processes (e.g., allergic rhinitis associated with nasal polyps). Treatment of specific diseases (e.g., correction of the postviral changes in taste and smell with zinc ion or receptor abnormalities following head injury or allergic rhinitis with theophylline) can be useful in restoring function in some patients. Other therapeutic measures for treatment of taste and smell dysfunction, including pyridoxine, vitamin A, adrenal cortical steroids, hydergine, alpha receptor agonists, local cromolyn sodium, dipyridamole, or treatment of dysosmia or dysgeusia with anticonvulsant agents (e.g., depekene) or butyrophenones can be useful either separately or in combination. Treatment with each drug has specific antecedents and consequences, and each requires special clinical management that may be different from that associated with more common uses of these drugs.

Prognosis

Treatment outcome depends upon several factors including severity of dysfunction and etiological factors, but not length of prior existence of the disorder. Treatment of PIHH with associated dysgeusia and dysosmia is commonly successful if hypogeusia is present. Treatment of hyposmia after head injury with theophylline is more successful if type II hyposmia is present but less so in the presence of type I hyposmia. On the other hand, treatment of patients with allergic rhinitis in the presence of type I hyposmia or anosmia is successful in four of five patients, whereas in the presence of type II hyposmia a lower success rate is achieved. In patients with congenital hyposmia with type I hyposmia or anosmia no more than 3 in 10 patients regain smell function. Treatment of dysosmia and dysgeusia is commonly successful, especially if the symptoms are isolated, unrelated to hypogeusia or hyposmia.

Mechanism of dysfunctions

Taste and smell are chemical senses intimately related to the biochemistry of the body. Even mild changes in well-being are known to induce anorexia and dysappreciation of tastants, vapors, foods, and beverages. Changes in the initial event of tastant or vapor binding to specific receptors depend upon the maintenance of adequate numbers and function of receptors. This is dependent upon specific proteins in the peripheral system that induce growth and provide nutrition for these receptors (for taste, gustin appears to be required, and for smell a gustin-like protein, i.e., a zinc metalloprotein). Transduction of both taste and smell information depends upon adenylate cyclase and the cascade of events that follows activation of cAMP. Calmodulin is also involved in this aspect of the system. Hormones, divalent cations, particularly calcium and zinc, drugs, and other substances influence this aspect of the system both positively and negatively. Transmission of taste and smell information along neural pathways depends upon the same axonal and synaptic factors that influence other sensory pathways; mechanisms by which integration of these sensory signals occur are unclear, as are the neural centers responsible for these events.

Pathological studies of patients with smell dysfunction have identified three important brain projection areas in humans including the subfrontal (orbital) cortex, the anterior portion of the temporal lobe, and the thalamus, including the medial dorsal nucleus. Excision of the anterior portion of the temporal lobe in patients with temporal lobe seizures results in gustatory and olfactory agnosias, particularly if some of the dominant hemisphere is affected. These pathological findings suggest a basis for understanding the mechanisms that underlie the normal and pathological function of molecular detection, recognition, and integration of liquid and vapor phase stimuli in these complex chemosensory systems.

Further reading

Schechter PJ, Henkin RI (1974): Abnormalities of taste and smell following head trauma. *J. Neurol Neurosurg Psychiat* 37:802–810

Henkin RI (1984): Zinc in taste function: a critical review. *Biol Tr El Res* 6:263–280

Mattes-Kulig DA, Henkin RI (1985): Energy and nutrient consumption of patients with dysgeusia. *J Am Diet Assoc* 85:822–826

Law JS, Henkin RI (1986): Low parotid saliva calmodulin in patients with taste and smell dysfunction. *Biochem Med Met Biol* 36:118–124

Henkin RI (1975): Effects of ACTH, adrenocorticosteroids and thyroid hormone on sensory function. In: *Anatomical Neuroendocrinology,* Stumpf WE, Grant LD, eds. Basel: Karger AG

Henkin RI (1976): Taste in man. In: *Scientific Foundations of Otolaryngology,* Harrison D, Hinchcliffe R, eds. London: Wm. Heinemann Medical Books

Henkin RI (1982): Olfaction in human disease. In: *Looseleaf Series of Otolaryngology,* English GM, ed. New York: Harper and Row

Taste Bud

Albert I. Farbman

Distribution

Taste buds are sense organs that respond to certain chemical stimuli presented to them in solution. In land-dwelling vertebrates, these bud-shaped organs are located almost exclusively in the mouth and pharynx, whereas in some aquatic animals, such as catfish, they are found on exterior surfaces of the body as well. In mammals, most taste buds are located within the epithelium of the fungiform, foliate, and circumvallate papillae on the tongue; smaller numbers of them are found in the soft palate, tonsillar pillars, pharynx, epiglottis, and larynx.

Morphology

The anatomy of taste buds is similar in most vertebrates. They are bud-shaped structures, usually 50–100 μm high and 30–60 μm in diameter, that are interspersed within the oral epithelium. They extend from the base of the epithelium to its surface. Each bud is made up of approximately 30 to 80 or more elongated, spindle-shaped epithelial cells. The nucleus is most often in the middle or lower third of the cell; the basal end rests on a basement membrane, and the apical portion tapers and ends in a small (2–5) μm opening or pore in the epithelial surface (Fig. 1).

In aquatic animals (e.g., fish, neotene amphibians) the opening in the epithelial surface is considerably broader (up to 25 μm diameter). The apical parts of taste bud cells end in microvilli or club-shaped processes which project into the pore and come into direct contact with oral fluids that contain taste stimuli.

Sensory nerve terminals enter the taste epithelium at the base of the bud, and end on the surfaces of the component cells, mostly in the basal and perinuclear region. These nerve terminals represent the primary innervation which carries the sensory message from the epithelial cells of the receptor organ to the second-order neurons in the brain stem.

An examination of the detailed histology of vertebrate taste buds reveals that it is similar, but not identical, from one species to another. Figure 1 is a low-magnification electron micrograph of a 0.5-μm thick section of a taste bud from a monkey foliate papilla, sectioned longitudinally and including the pore region. Taste buds from this region contain three cell types, characterized on the basis of their morphology (Farbman et al., 1985). They are called, simply, type I, II, and III cells. Approximately 60% of the cells are type I, about 30–35% type II, and the remainder type III. The type III cell in the monkey foliate papilla taste bud is thought to be a receptor cell because its relationship to the intraepithelial nerves is similar to that of a synapse. The type I cell is thought to be a secretory cell, responsible for secretions into the pore region which may play a role in maintaining a constant micro-environment for the taste cell microvilli. The function of type II cells is unknown.

Most recent work has shown that taste buds from some other regions, e.g., foliate papillae of rabbits or human fungiform papillae, also contain three cell types. However, some taste buds, e.g., those from rat fungiform papilla, mouse circumvallate papilla, and monkey larynx (fetal), contain only two. In these, the homolog of the type III cell seems to be absent, and it is thought that type II cells, or both I and II may be chemosensory (Ide and Munger, 1980).

Based on the evidence now available, it is fair to say that although the anatomy of taste buds in vertebrates is fairly similar, some variation does occur. No association between the different cell types and specific taste sensitivity has ever been demonstrated, i.e., it has not been possible to associate salt taste with one of the cell types, sweet taste with another, etc. (Tateda and Beidler, 1964).

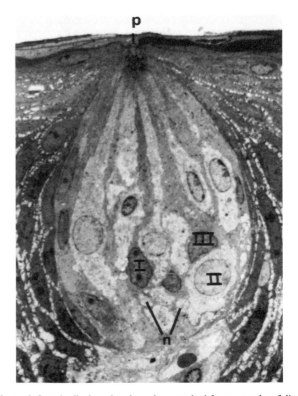

Figure 1. Longitudinal section through a taste bud from a monkey foliate papilla. The taste pore (p) is shown at the apex where the cells taper to end in microvilli. Several nerves (n) are shown near the base of the bud. The three cell types, I, II, and III, are seen distinctly.

Plasticity

One interesting aspect of taste bud biology is that the constituent cells have a finite life span and are continuously replaced. Type I cells turn over every 9–10 days. The other cells have longer lives, but there are no precise data on the length of their life cycle (Farbman, 1980). This continuous cell renewal has several important implications with respect to taste function. Under physiological conditions, cell renewal does not occur synchronously; therefore, at any given time in an adult taste bud, old and young cells of each of the types are likely to be present. If young and old cells respond to tastants differently, as is suggested by some recent experiments (Hill et al., 1982), the response spectrum of a given taste bud would depend on the relative ages of its cells. For example, if, at a particular time there were, by chance, a preponderance of young, immature cells in a given taste bud, the response pattern might differ from what it would be at a later time when these cells become mature. Furthermore, if cells are continuously replaced, new cells would constantly have to form synaptic relationships with sensory nerve terminals as they become functional. These biological phenomena must be taken into account in any consideration of the mechanism of taste perception.

Physiology

There is a considerable body of evidence on the physiology of taste buds, but the mechanism of taste transduction remains a mystery. Early physiologists identified four basic taste modalities, namely salt, sour, sweet, and bitter, and these have been the convenient stimulus references for physiologists and psychologists studying taste. The bulk of the existing evidence suggests that neither individual cells within a taste bud nor individual taste nerve fibers are responsive to only one taste modality; rather, it seems that they are usually responsive to two or more. Although the notion of four basic modalities is a convenient one for some experimental purposes, most substances that elicit a taste response produce a sensation that cannot be simply characterized. Some investigators have suggested that there may be other basic taste modalities, such as alkaline, metallic, and pungent. However, there is now reason to believe that taste modalities may really not be distinct but may be part of a complex stimulus continuum (Schiffman and Erickson, 1980).

Further reading

Farbman AI (1980): Renewal of taste bud cells in rat circumvallate papillae. *Cell Tiss Kinet* 13:349–357

Farbman AI, Hellekant G, Nelson A (1985): Structure of taste buds in foliate papillae of the rhesus monkey, Macaca mulatta. *Am J Anat* 172:41–56

Hill DL, Mistretta CM, Bradley RM (1982): Developmental changes in taste response characteristics of rat single chorda tympani figers. *J Neurosci* 2:782–790

Ide C, Munger BM (1980): The cytologic composition of primate laryngeal chemosensory corpuscles. *Amer J Anat* 158:193–209

Schiffman SS, Erickson RP (1980): The issue of primary tastes versus a taste continuum. *Nuerosci Biobehav Rev* 4:109–117

Tateda H, Beidler LM (1964): The receptor potential of the taste cell of the rat. *J Gen Physiol* 47:479–486

Temperature Regulation

James M. Lipton

Many living organisms maintain internal temperature within narrow limits in spite of changing internal and external conditions that tend to cause it to deviate. Humans and other higher vertebrates are homeotherms, organisms capable of regulating core temperature within narrow limits through physiological (e.g., sweating, shivering) and behavioral changes driven by thermal discomfort. Poikilothermic animals such as reptiles exhibit greater variation in temperature and primarily use behavioral (e.g., voluntarily moving to a more temperate environment), rather than less well-developed physiological, thermoregulatory responses. With the exception of sweating in humans, heat loss and heat production effector mechanisms serve other functions as well (e.g., muscles used in shivering are used for voluntary movement), and in both homeotherms and poikilotherms strict adherence to temperature regulation can be set aside for brief periods in favor of other activities. The temperature of circulating blood in humans is not homogeneous, and deep body temperatures vary with the site of measurement. For example, in humans rectal temperature tends to be 0.5–0.7°C higher than oral and axillary temperatures. The temperature of normal persons is not necessarily 37°C at any particular measurement site, and there is considerable variability in the range of temperatures among individuals. Some people have average temperatures near the upper and lower limits of 36–38°C.

Temperature increases can lead to heatstroke and are generally more dangerous than decreases. Hypothermia can also lead to death, but in controlled conditions (e.g., surgery) reduced body temperature may be useful because of reduced metabolic demand. It is possible to successfully revive a laboratory animal that has been cooled below the level of cardiac and respiratory arrest to a temperature at which a significant portion of the body water is ice.

Body temperature is regulated around a set temperature. It is generally maintained within a band or range that is associated with low or negligible physiological heat production or heat loss activity. Human set temperature varies according to a circadian rhythm (0.7–1.5°C range in adults) believed to be determined by a pacemaker within the hypothalamus and the menstrual cycle, in which temperature increases after ovulation and decreases after menses. Human neonates have less precise temperature regulation because the temperature control system has not yet fully developed. Thermoregulatory functions, particularly the capacities to develop fever, to respond rapidly and adequately with thermoeffector activity, and to sense extreme ambient temperatures as uncomfortable, are decreased in the elderly, in major part due to changes in central nervous system temperature controls.

The control system

Temperature regulation depends upon central and peripheral nervous structures that detect impending or actual changes in deep body temperature and drive appropriate counter responses. Cutaneous thermoceptors are unencapsulated nerve endings that respond solely or preferentially to temperature. Receptors that respond to cold are nearer the surface of the skin than those that respond to heat. Activity of the receptors is presumed to provide discrete sensory information via lemniscal pathways to the thalamus and to influence central temperature controls via reticular pathways. Temperature sensors also reside within the abdominal cavity and the brain.

The primary temperature control in higher organisms is based within the preoptic/anterior hypothalamic (PO/AH) region of the brain. In this region as many as one in five neurons is sensitive to temperature change. Whereas all cells are affected by temperature, these neurons respond with $Q_{10} > 2$ over the physiological range of temperature. These neurons also receive input from thermosensitive elements in other parts of the nervous system and from the periphery, providing for anticipatory thermoeffector changes before there are changes in deep body temperature. Thermosensitive neurons are also distributed, perhaps less densely, in the midbrain, pons, medulla, and the spinal cord.

Two general categories of thermosensitive neurons have been described: warm-sensitive units increase firing rate with increases in temperature, and cold-sensitive units decrease firing rate with decreases in temperature. Whereas the heat-sensitive neurons are believed to be true "direct" thermoceptors, there is a question as to whether cold sensitivity reflects interneuron status and inhibition by warm-sensitive neurons. Thermal stimulation of the PO/AH region, the medulla oblongata, or spinal cord evokes physiological and behavioral thermoregulatory responses that shift core temperature in the opposite direction (e.g., local cooling causes shivering and a rise in core temperature above resting or prestimulation temperature). The influence of medullary temperature on behavioral thermoregulation is enhanced after PO/AH destruction, indicative of an increased importance of lower circuits to the type of thermoregulation used by precursors of the modern homeotherms. There are differences in precision of control exerted by these three thermosensitive regions that suggest a hierarchy of temperature controls. Rapid destruction of the PO/AH region results in less precise or broad-band temperature control, presumably the level of temperature regulation afforded by the remaining medullary and spinal cord controls.

Clinically, a number of patterns of dysthermia have been observed with CNS lesions: relative thermolability (broad-band control), deficits in specific thermoeffector activity (e.g., sweating) that result in less precise control, deficits in response to heat or to cold alone, etc. Children with CNS lesions are more likely to have increased temperatures. In contrast to fever caused by infection, these lesion-related hyperthermias are not antagonized by antipyretic drugs. Adults with dysthermia due to CNS injury are more likely to be hypothermic since the most common disorder is thermolability and the temperature of most habitats is below body temperature.

A number of neurotransmitters including catecholamines, serotonin, and amino acids such as taurine, serine, and glycine have been implicated in CNS control of body temperature. One prominent theory is that central catecholamines are important to heat loss and serotonin to heat production. This theory, developed using subhuman primate and cat models, appears not to hold for all species. A central ionic theory of temperature control states that the temperature setpoint depends upon the sodium-to-calcium-ion ratio within the posterior hypothalamus. This concept is supported by evidence of local changes in ionic ratio during thermal challenge and fever in several species.

Fever, and hyperthermia in anesthesia

Fever is an elevation in the level around which body temperature is regulated. This change occurs when endogenous pyrogen (EP) or interlenkin 1, a product of the interaction of host cells and microbial invasion or an injury, alters the firing rate of thermosensitive neurons, preferentially increasing the firing of cold-sensitive neurons and decreasing that of heat-sensitive cells. The precise characteristics of the EP that enters the brain are not known. One fragment of about 14 kilodaltons has been discovered, but it is likely that much smaller fragments actually penetrate to the temperature controls. Whatever the nature of EP, the fever-producing capacity of experimentally produced EP may be universal since human EP causes fever in monkeys and rabbits, cat EP causes fever in monkeys, and so on. There may be neurochemical intermediates, such as arachidonic acid metabolites (e.g., PGE) or protein mediators, between EP and temperature controls, but the existence of such mediators has not been established with certainty. Fever rarely rises above 41.1°C, perhaps because of central release of fever-reducing peptides related to adrenocorticotropin, such as melanotropin (α-melanocyte stimulating hormone). This latter peptide is more than 25,000 times more potent than acetaminophen in reducing fever in the rabbit. The upper limit of fever is guarded closely and cannot be raised by cooling the medulla oblongata or the PO/AH region in primates. On the other hand, warming either region causes rapid reduction in fever. Defervescence depends upon a transport process within the brain, presumably essential to central inactivation of EP, that can be inhibited by taurine or related molecules and by transport inhibitors such as probenecid. Antipyretic drugs may act centrally to release endogenous antipyretic substances or to compete with EP for receptor sites. Most antipyretics do not reduce normal temperature, and there is evidence that they have no effect on thermoresponsiveness of the PO/AH and medullary controls in squirrel monkeys. Fever mediation appears to be organized at very low levels of the nervous system, and there is evidence of fever development in fish and lizards via changes in behavioral thermoregulation. Although the PO/AH region is the most sensitive part of the brain in terms of response to injected pyrogens, fever capacity has been shown to persist in humans after naturally occurring lesions and in monkeys after experimental destruction of the PO/AH.

Marked hyperthermia (malignant hyperthermia) results in some persons and animals when they are given anesthetic or are stressed. The primary problem is massive heat production due to increased muscle tone traceable to an inborn error of calcium metabolism. In anesthesia, the temperature control mechanisms are temporarily unable to evoke appropriate compensatory thermoeffector actions, and the patient may succumb to marked hyperthermia.

Aging

With aging there are both increases and decreases in central sensitivity to substances that alter temperature in younger organisms. There is a reduction in fever capacity in older persons and other homeotherms that has been traced to a reduction in central sensitivity to endogenous pyrogen in experiments on monkeys and rabbits. Many centrally acting drugs commonly taken by older persons, such as alcohol, chlorpromazine, and diazepam, have greater dysthermic effects via actions on the aged brain. The aged brain is also more sensitive to neurotransmitter substances, such as α-melanocyte stimulating hormone, taurine and β-endorphin, that markedly alter body temperature.

Further reading

Clark WG, Lipton JM (1983): Brain and pituitary peptides in thermoregulation. *Pharmacol Ther* 22:249–297

Hardy JD (1980): Body temperature regulation. In: *Medical Physiology*, Mountcastle VB, ed. St Louis: Mosby

Lipton JM (1984): Thermoregulation in pathological states. In: *Heat Transfer in Biological Systems: Analysis and Application*, Shitzer A, Eberhart RC, eds. New York: Plenum Press

Myers RD (1984): Neurochemistry of thermoregulation. *Physiologist* 27:41–46

Thermoreceptors

David C. Spray

Thermoreceptors are of several types. Those in the hypothalamus, spinal cord, and gut apparently monitor body temperature and are involved in autonomic functions. Those in the skin, the cutaneous temperature receptors, mediate localized and general temperature sensations in humans and participate in the autonomic processes achieving endothermy in mammals and birds and in behavioral thermoregulation in submammalian species. The sensory receptors that transduce maintained and transient skin temperatures are defined by their exclusive or low threshold excitation by cold or warm stimuli. The impulse frequency recorded from a thermoreceptor afferent is consistently related to maintained skin temperature and is transiently altered by changing skin temperature. Threshold and sensitivity are comparable to those psychophysically determined in human subjects and established in studies of animal behavior. Other receptors are affected by temperature, particularly the slowly adapting mechanoreceptors, and it is still unclear whether they are also involved in either perception of heat and cold or processing of thermal information.

Cold receptors are distinct from warm receptors, and responses to both, often in the same nerve innervating the same skin area, have been characterized in humans and other primates, as well as in cat, dog, rat, hamster, and frog (sensory surfaces examined include both hairy and glabrous skins of face, tongue, scrotum, and extremities). For maintained temperatures in the range from 20° or lower to 30°C or higher, the curve relating impulse frequency of cold receptors to skin temperature is unimodal, with a peak at about 25–27°C (Fig. 1).

Cold receptors also respond to very high temperatures, starting above about 45°C; this activation is correlated with the paradoxical cold sensation that humans experience when the skin is exposed to such high temperatures. For warm receptors, discharge is initiated at temperatures above 30°C. Activity increases as temperature rises, reaching maximum discharge at 41–47°C, and then decreases, often abruptly, as the skin

is warmed further (Fig. 1). The response curves of warm and cold receptors overlap in the range 32–35°C. In some preparations, thermoreceptor activity at sustained skin temperatures is random with temperature-dependent average rates, others show slow fluctuations, and in other preparations, particularly in cold receptors, impulses occur as couplets or groups (''bursts''). The bursting in cold receptors generally appears as temperature is reduced below the temperature of peak activity and may serve to resolve the neural code which is redundant for maintained temperatures above and below the peak discharge of the cold receptor.

Thermoreceptors respond to temperature changes with transiently increased or decreased activity (cold receptor activity increases when the skin is cooled and decreases when warmed whereas warm receptor activity is conversely affected) that then relaxes to the activity rate appropriate for the steady temperature. The kinetics of the dynamic phase are apparently simple, with time constants that depend on magnitude and rate of temperature change. The dynamic response of temperature receptors has as its psychophysical correlate heightened perception of changes in skin temperature; thermal sensation is a graded function of both magnitude and rate of thermal stimulation.

Free nerve endings located in the dermal and epidermal skin layers are the apparent terminations of the fibers from which specific thermal responses arise. The actual sensory receptors are apparently mitochondria-rich conical or bulbous projections of nerve fibers into the cytoplasm of basal epithelial cells. These projections are branched unmyelinated processes from afferent fibers which are commonly dichotomized into the A delta category of myelinated axons for cold receptors and C fibers for warm receptors. This separation is consistent with differential sensitivity of the two receptor types to anesthesia but is not absolute; smaller fractions of cold receptors are found among afferents with C-fiber conduction velocities and of warm receptors with myelinated axons. Specific thermore-

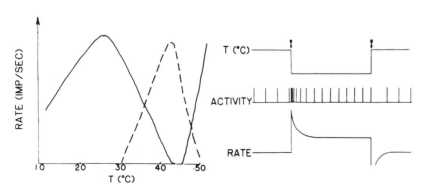

Figure 1. Relation of nonadapting activity to maintained skin temperature for typical mammalian cold (solid line) and warm (dashed line) receptors. The response curve for cold receptors at sustained skin temperatures shows a peak at about 25–27°C; activity is also increased at very high temperatures corresponding to the sensation of paradoxical cold. Warm receptor activity peaks at about 45°C. The receptor response curves cross, with low rates of activity, in the range of normal skin temperature.

ceptor axons are smaller than those of temperature sensitive mechanoreceptors, which are recruited into the afferent response at higher stimulus strengths; the result is an afferent fiber population whose mean conduction velocity increases with increasing stimulus strength.

The branching that occurs as free nerve endings penetrate the dermis gives rise to punctate or oblate receptive fields for single fibers, generally about 1 mm^2 or less in diameter. Larger and multiple fields are sometimes found and can allow spatial and temporal summation of thermal information.

Primary thermal afferents from skin receptors have somata in representative sensory ganglia and project to the marginal zone of the spinal cord, synapsing on rostral brain stem and thalamic nuclei and projecting to somatosensory cortex. Integration at the first level of synaptic interaction is apparently considerable; the representation in the dorsal horn can be large and bilateral and there may even be convergence of warm and cold thermoreceptors and possibly temperature-sensitive mechanoreceptors. The brain stem areas involve several midbrain raphe nuclei; thalamic areas of thermoreceptor representation include VB and NPT. Relay to the hypothalamus, where there is integration with information from deep body thermoreceptors, is apparently from midbrain raphe nuclei parallel to the thalamic projection.

Descending control of thermosensory information is largely at or before the initial synapse of the primary afferent, in contrast to supraspinal control of the pain pathway. Segmental modulation includes convergence and also effects of mechanical and sympathetic stimulation that apparently act directly on the receptor. Psychophysical correlates include "Weber's deception," in which cold objects feel heavier than neutral ones and, possibly, temporal variations in thermal thresholds.

The transduction mechanism of thermoreceptors presumably involves temperature sensitivity of a number of neuronal properties, including ionic conductances and pumps in the end organ. A metabolic transduction mechanism largely accounts for excitation of cold receptors. The hypothesis is that the cold receptor transduces by nature of the high surface area to volume of its terminal process and activity of ionic pumps in terminal membranes. Operation of the highly temperature-sensitive sodium-potassium pump would maintain the resting potential of the thermoreceptor membrane, so that a drop in temperature would decrease pump activity and thus depolarize the terminal, giving rise to receptor current. The pump could be electrogenic and thus contribute directly to the resting potential by generating current across the membrane resistance. Or the pump contribution might be nonelectrogenic, controlling gradients of permeable ions and thus altering membrane potential as predicted by the Nernst-Goldman equations. With either sort of pump, the effect of decreased temperature would be the dissipation of resting potential, due either to cessation of pump current or collapse of ionic gradients.

The degree to which nonmetabolic temperature-dependent processes account for excitation of thermoreceptors is unclear. Because the response of cold receptors is bell shaped (Fig. 1), an inhibitory process may operate at lower temperatures. Also at low temperatures many cold fibers exhibit burst discharge, suggesting that there may be an underlying oscillation in generator potential. Adequate description of warm receptor transduction clearly requires another underlying process, one of which may be the asymmetric temperature dependences of Na and K permeabilities that participate in the transduction of cold thermal sensations. Alternatively, membranes of warm receptors might be very leaky so that the electrogenic pump would necessarily be stimulated to reactivate the ending.

Further reading

Hensel H (1981): *Thermoreception and Temperature Regulation.* New York: Academic Press

Schmidt RF (1983): Thermoreception. In: *Human Physiology,* Schmidt RF, Thews G, eds. New York: Springer-Verlag

Spray DC (1986): Cutaneous temperature receptors. *Annu Rev Physiol* 48:625–38.

Time Perception

Ernst Pöppel

Immanuel Kant once said: "A science has as much truth as it has mathematics." This remark puts psychology in a bad position, but the investigation of human time perception is probably in an even worse position as far as scientific level is concerned—there is certainly no mathematics in it because there is no consensus about what time perception actually is. What is needed is a classification system or a taxonomy of temporal experience. As long as there is no acceptable classification, the application of mathematical reasoning is precluded. A scientific analysis of temporal experiences requires the development of a taxonomy in order to prepare the ground for more theoretical work.

Time perception can perhaps best be understood by using a hierarchical system in which different elementary temporal experiences (ETEs) are incorporated at different levels. At least four such ETEs can be identified: the ETE of simultaneity or nonsimultaneity; the ETE of temporal order or succession; the ETE of "now" or the subjective present; and the ETE of duration. Although the ETEs may appear as independent phenomena of subjective reality, an internal structure, or hierarchy, exists that relates the different ETEs to each other.

The most basic ETE is our ability to discriminate between simultaneity and nonsimultaneity of stimuli. Different sensory systems have different thresholds for the nonsimultaneity of stimuli, which are determined by measuring the fusion threshold. This highest temporal resolution (the lowest threshold) is observed in the auditory system. With binaural click stimulation, fusion threshold is approximately 2–3 msec. The somatosensory system has a slightly higher threshold. But the highest threshold with respect to temporal resolution is found in the visual system, which is at least 10 times less sensitive than the auditory system. The fact that the fusion thresholds of the three modalities differ indicates that they might each be dependent on mechanisms at their receptor level. Each sensory system has its own transduction mechanism, and the time taken for transduction appears to be responsible for the temporal resolution of the system.

It is interesting that the detection of nonsimultaneity of stimuli is not sufficient to indicate their temporal order. If a subject hears two clicks with a 5-msec delay between them, he may be quite certain that he heard two, but he cannot tell which one came first. It takes approximately 30 msec to identify the temporal order of auditory events. If order threshold is measured in the visual or somatosensory system, a value of 30 msec is also obtained (depending on conditions that may vary from 20 to 50 msec). Thus, fusion threshold and order threshold are not correlated. Nonsimultaneity of stimuli is necessary but not sufficient to identify order. The ETE of succession or temporal order has a higher position in the hierarchical taxonomy of time perception because there is an additional mechanism at work. While fusion threshold for consecutive stimuli appears to be dependent on transduction at the periph-

eral level, order is apparently centrally mediated. This hypothesis is based on the observation that order threshold is the same for the three sense modalities tested, suggesting a common central mechanism. Further support comes from observations on patients who have suffered an aphasia after a left hemisphere lesion. Such patients show auditory order thresholds of almost 100 msec, whereas their auditory fusion threshold shows normal 3-msec values or is only changed minimally, indicating the central basis of auditory temporal order perception.

The order threshold of 30 msec for successive events can also be considered as the temporal threshold for the identification of an event. At the next level of this hierarchical taxonomy, there is again an additional mechanism (an integration) that binds separate successive events together into a unit. The temporal extent of this integration is limited on the average to 2 to 3 sec. The ETE that is mediated by this integration is usually referred to as the subjective present or the feeling of "now." What is the experimental basis for assuming an integration interval of only a few seconds? Evidence comes from different areas. If the duration of temporal intervals has to be reproduced, subjects overestimate these intervals by a small amount up to approximately 2 to 3 sec; they underestimate longer intervals usually substantially. The indifference interval between over- and underestimation has been interpreted as the temporal limit of integration. A similar indifference interval is observed when stimulus intensities (like loudness or brightness) in different sense modalities have to be reproduced, and the delay between stimulus presentation and reproduction is varied. Subjects overestimate time up to the indifference interval and underestimate it beyond. It appears as if the sensory systems can hold information only up to the temporal limit of 2 to 3 sec and that longer delays result in a representational leakage.

There is additional evidence from completely different experimental paradigms that also suggests a categorical border with respect to our handling of time. For example, it has been shown that Weber's law ($\Delta S/S$ = constant; S representing stimulus duration) holds only up to a temporal limit of a few seconds. Some researchers thus talk about time perception only for intervals within a range of 2 to 3 sec, and use the term time estimation for longer intervals.

Some other observations can also be related to the hypothesis of an integration with a limit of 2 to 3 sec. Spontaneous speech in different languages (like English, Spanish, German, or Chinese) is organized in such a way that verbal utterances usually last 2 to 3 sec and are followed then by a pause that is used to organize the next utterance. Apparently, the integration mechanism does not allow for anticipatory planning exceeding this temporal limit. Another example comes from the spontaneous rate of alteration of ambiguous figures. For instance, it has been found that each perspective of the Necker cube, if allowed to switch spontaneously, lasts only 2 to 3 sec on average. These data suggest an integration that has a temporal limit of only a few seconds. Subjectively, this integration is suggested to be the basis for what we experience as ''present'' or ''now.''

Whereas physiological mechanisms may explain the ETEs of nonsimultaneity, order, or the subjective present, no such physiological explanation is in sight for the ETE of duration. It has been shown that duration is mainly dependent on how much information has been processed in a given interval. If mental content was high, the duration of that interval appears to be long; if mental content was low, the duration appears to be short. But this refers only to a retrospective evaluation of time. If one is in a situation with low informational content (if one is bored), time appears to slow down. If mental content is high because of a lot of interesting information, time appears to pass very quickly. Thus, time past and time present give us completely different views of the ETE of duration. This contradictory experience is usually referred to as temporal paradox.

The ETE of duration is at the highest level of the hierarchical taxonomy of time perception. In order to experience duration, an additional mechanism comes into play, i.e., a mnemonic device. Without memory, there would be no duration. And, in fact, studies with patients who have lost their memory due to surgical intervention or brain trauma show a disturbing shrinkage of subjective duration.

Further reading

Cohen J (1967): *Psychological Time in Health and Disease*. Springfield, Ill: CC Thomas

Fraisse P (1963): *The Psychology of Time*. New York: Harper & Row

Ornstein RE (1969): *On the Experience of Time*. New York: Penguin, Harmondsworth

Pöppel E (1978): *Time Perception. Handbook of Sensory Physiology Vol 8. Perception*. Held R, Leibowitz HW, Teuber HL, eds. Berlin: Springer-Verlag

Pöppel E (1985): *Grenzen des Bewusstseins. Über Wirklichkeit und Welterfahrung*. Stuttgart: Deutsche Verlagsanstalt

Touch, Sensory Coding of, in the Human Hand

Å.B. Vallbo

The sense of touch has its most significant role in the human hand where it is essential for exploration and manipulation. These two activities are based on a close integration of motor and sensory activities and are of paramount importance to the human being from childhood when the physical world is first explored and a model of it is constructed within the brain through adulthood when the hand is used as a powerful, versatile, and accurate instrument, "the cutting edge of the brain" as Jacob Bronowski emphasized in his famous series, *The Ascent of Man.*

Although several kinds of sensory mechanisms, including muscle sense and kinesthesia, are essential for advanced hand functions, it is evident that the mechanosensitive receptors in the glabrous skin play a key role.

New experimental approaches have recently been developed that provide insight into tactile functions of the hand. These methods include microneurography recording from single nerve fibers and microstimulation of single fibers while the subject is fully awake and attending to a psychophysical test task.

The glabrous skin of one hand is equipped with about 17,000 sensory units that are sensitive to nonnoxious mechanical deformation of the skin. They constitute the peripheral basis of the enormous capacity for spatial and temporal discrimination in this skin area. A sensory unit consists of a nerve cell with its long axon running in the peripheral nerve and the associated end organ or end organs in the skin or the subcutaneous tissues.

The 17,000 units are of four types with distinctly different response properties. Two of them are fast adapting and two are slowly adapting to sustained indentations. The fast adapting type I units (FA I, earlier called RA or QA) are very likely connected to Meissner corpuscles. There are at least 10–20 of them scattered within an area of 2–20 mm^2. The FA I units respond exclusively to moving stimuli and only when skin deformation occurs close to the endings. A similar organization is exhibited by the slowly adapting type I units (SA I), which are connected to Merkel cells located in the basal layer of the epidermis. The terminals are distributed within a small area of similar size, as with the Meissner units. They exhibit a high dynamic sensitivity to moving stimuli but respond to stationary stimuli as well.

These two kinds of units constitute about 70% of the tactile units in the glabrous skin. Characteristically they occur in high densities at the fingertips. There are about 140 FA I units and 70 SA I units per cm^2 at the fingertip, whereas the density is 5–10 times lower in the palm (Fig. 1A). A number of factors point to these units being essential for spatial discrimination. Their small receptive fields with distinct borders make up a multiple overlapping array that constitutes a fine-grain system allowing accurate localization of stimuli. The ability of two-point discrimination as analyzed in psychophysical tests with human subjects varies between subregions of the hand in close relation to the density of these two kinds of units (Fig. 1A) and, in the fingertip the two-point discrimination is actually close to the theoretical limit of this organization.

It is interesting that the distinct receptive field of these units as defined in recordings of impulses in the afferent nerve fiber has a counterpart in the perceptive space of the mind. When the microstimulation method is used to inject a train of impulses through an identified afferent fiber of an attending human subject, a percept is produced that often is characterized by identical spatial properties as those of the receptive field. The percept is accurately localized to the identical skin area and its extent on the skin has the same size and shape as the receptive field of the afferent unit, indicating that an individual afferent unit may be accurately represented within the brain.

In addition to type I units (FA I and SA I) there are two kinds of units that have distinctly different functional properties from those of the Meissner and Merkel units. The fast-adapting type II units (FA II, earlier PC) are presumably connected to Pacinian corpuscles and related kinds of smaller lamellated endings denoted Golgi-Mazzoni bodies. The end corpuscles of these units are actually not located in the skin but in the subcutaneous tissue. Still, the sensitivity of the FA II units for skin deformation is extremely high, particularly for rapidly moving stimuli. The slowly adapting type II units (SA II) are presumably connected to the spindle-shaped Ruffini endings.

The afferent fibers of type II units terminate in connection with a single end organ in contrast to the multiple endings of type I units. On the other hand, type II units are sensitive to remote stimuli with the effect that a localized deformation will excite a number of units having endings spread over a large skin area. Hence, capacity of these systems to define accurately the localization of a stimulus is low. Moreover, the density of type II units within different skin regions of the hand is unrelated to the spatial discrimination capacity of human observers in psychophysical tests (Fig. 1A).

Generally the density of FA II and SA II is much lower than for the type I units and fairly uniform within the whole glabrous skin area, from the fingertips to the wrist. All these characteristics support the interpretation that their main role is not in spatial analysis of stimuli but in extraction of other features of the tactile stimuli.

The Ruffini units (SA II) have the unique property to respond not only to direct skin indentations but also to stretching of the skin which normally occurs during movements of the joints. On this basis, it has been suggested that they have a significant role in kinesthesia. Moreover, in manipulation of objects with the hand, the Ruffini units would respond to the tangential forces in the skin and might provide information of significance for controlling the grip force to avoid slipping. When a slip occurs, however, the other units respond and elicit a reflex response from the muscle to counteract dropping the object.

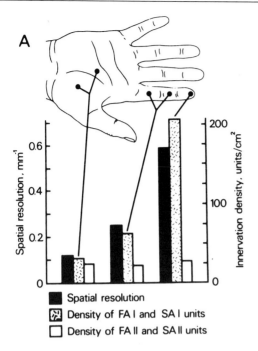

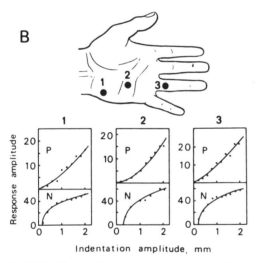

Analysis of single afferent's response in relation to detection of touch stimuli in the glabrous skin of the hand demonstrated that one or the other threshold concept is not universally valid. In contrast, the limiting factor varies depending on which skin region is stimulated. When stimuli are applied with a small probe, the touch threshold of human observers is identical with the threshold of the most sensitive units (FA I and FA II) on the volar aspects of the fingers, with the exception of the creases and periphery of the palm (Fig. 2). Actually, a single impulse in a Meissner unit (FA I) may elicit a detectable touch sensation when the subject was maximally attending. This was not true for the FA II units. This finding was confirmed with microstimulation. A single electrical pulse delivered through the microelectrode to excite a single and identified FA I unit often elicited a noticeable touch sensation on the skin area where the endings of the unit were located.

In other skin regions, particularly in the center of the palm, the psychophysical threshold was higher and above the threshold of the most sensitive afferent units (Fig. 2), indicating that central mechanisms set the capacity of detection. Moreover, there were indications that the level of noise was substantial in the parts of the somatosensory system concerned with these skin regions.

Figure 1. A. Spatial resolution capacity in psychophysical tests related to density of tactile units in three glabrous skin regions: fingertips, rest of the finger, and palm. Spatial resolution was defined as the inverse of the two-point threshold in units of 1/mm. From Vallbo and Johansson (1984), with permission. B. Comparison between neural stimulus-response functions (N) and psychophysical magnitude-estimation functions (P). Data from one subject extracted at three different test points (1,2,3) located in the center of the receptive fields of three SA I units which provided the neural functions (N). From Knibestöl and Vallbo (1980), with permission.

Several issues in the area of psychoneural correlations have been addressed by recent studies. The nature of the detection threshold is a long-standing issue in psychophysics and neurophysiology. Two groups of theories have been advanced. One group claims that the limit of detection of weak stimuli in psychophysical tests is set by the sensitivity of the sense organs in the periphery, whereas the other group claims that central mechanisms within the brain, such as decision strategy and noise, set the limit even when the subject is maximally attending to the task.

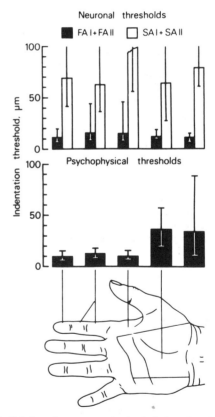

Figure 2. Relations between psychophysical detection threshold and thresholds of tactile units. Upper histogram shows the threshold for a single impulse in a sample of 128 units in various skin regions. Lower histogram shows psychophysical detection thresholds at corresponding skin regions to identical stimuli (162 test points). From left to right the columns give data from the terminal phalanx, the rest of the finger, the peripheral part of the palm, the central part of the palm, and to the extreme right, data from the lateral aspects of the fingers and the region of the creases taken together. Medians and 25th and 75th percentiles. Modified from Johansson and Vallbo (1979).

The analysis of detection capacity for touch in the human hand indicates that the nature of sensory threshold is not uniform but varies with stimulus location. For the fingertips, sometimes called the macula of the somatosensory system, and other important tactile regions, the threshold is set by the sensitivity of skin receptors, implying that the brain extracts information adequate for the secure production of a percept out of a neural quantum in the peripheral nerve. It is a corollary that neural noise is negligible at central levels of the somatosensory system in those parts which handle tactile information from these skin regions. In other regions of the hand, the threshold is set by mechanisms within the brain. Neural noise as implied in the signal detection theory might be one of the factors involved.

On the other hand, modern sensory physiology has advanced to a stage that makes the term sensory threshold or absolute detection threshold an unduly blunt concept without further specification. A variety of stimulus parameters such as velocity of movement, preindentation, and size of skin area indented influence the response from the population of units, notably which units are preferentially activated. Production of a sensation of touch depends on whether the message from the excited units reaches the perceptive levels.

Microstimulation makes it possible to analyze significant aspects of the functional properties of neural connections from single afferent units to the human perception. It has been shown that there is considerable variation among tactile units. A single impulse from a single Meissner unit (FA I) may produce a sensation of touch, whereas a train of 10–20 impulses from a Merkel unit (SAI) may be required.

Threshold stimuli are usually associated with sensations that are vague and not well specified except for approximate location and general quality. On the other hand, a number of attributes may be attached to sensations associated with stronger stimuli. Although it is generally accepted that sensory attributes are based on particular components of the composite afferent signal, it is a difficult problem to define which components are decisive for a particular attribute. The intensity of a sensation may, for instance, be based on impulse rate or number of units activated and other quantities as well.

In the sensation of touch it has been claimed that perceived intensity of a sustained indentation in the hand is set by the impulse rate in SA units. It has even been suggested that there is a linear relationship between their impulse rate and the perceived intensity, implying that the psychophysical function describing this relationship is determined at the receptor level and transformations at more central levels of the nervous system are in sum linear. These conclusions are based on collateral data from humans and monkey.

This problem was explored in humans when impulse discharge in single SA units were recorded while the subject made magnitude estimation of perceived intensity of rectangular indentations. Striking discrepancies were often found between the shape of the neural and psychophysical stimulus-response functions (Fig. 1B). Moreover, the relationship between the shape of the two functions was practically random in the total sample. These findings indicate that, if the intensity of the stimuli is coded in the impulse rate of SA units, the perceived intensity function of the human observer is shaped by the central nervous system rather than at the receptor level. Moreover, it seems that the individual brain produces consistently uniform intensity functions that vary among subjects.

It would be interesting to explore further the general problem of intensity coding using the microstimulation approach. Unfortunately, so far only casual observations on this point have been reported and do not justify any conclusions.

For the study of sensory coding in the somatosensory system of human subjects the microneurographic method offers the possibilities of correlative psychoneural studies on several levels. (1) Group data on functional properties of human mechanoreceptive units may be correlated with psychophysical data on tactile sensibility. (2) The response properties of a particular unit or sets of units may be defined and correlated with psychophysical responses of the subject whose afferents are recorded using identical stimuli with regard to location and stimulus parameters. (3) The signal in the somatosensory system may even be tapped simultaneously at two levels, in the peripheral nerve and as the psychophysical response of the attending subject when suitably designed stimuli are delivered. (4) With electrical microstimulation of identified single units, the subject's percept associated with a train of impulses in a single afferent channel may be analyzed to elucidate the functional properties of the connections from the separate kinds of afferent units to the perceptive level of the human brain.

In recent years a number of new approaches to the study of the somatosensory system have appeared, including brain scanning and powerful neuromorphological techniques. A wealth of new methods provide data which are mutually fertilizing and stimulating. One obvious trend, in human as well as subhuman species, is to combine neurophysiological and behavioral methods in the study of the somatosensory system at various levels from the peripheral nerve up to the cortex in attending subjects. Although the goal of arriving at a full understanding of the mechanisms used by the brain to produce purposeful percepts out of the enormous amount of information from our sense organs may still be far away, knowledge is increasing rapidly, and steps have been taken toward a much deeper insight, particularly with regard to the higher levels of the system.

Further reading

Johansson RS, Vallbo ÅB (1979): Detection of tactile stimuli. Thresholds of afferent units related to psychophysical thresholds in the human hand. *J Physiol* (*Lond*) 297:405–422

Knibestöl M, Vallbo ÅB (1980): Intensity of sensation related to activity of slowly adapting mechanoreceptive units in the human hand. *J Physiol* (*Lond*) 300:251–267

Vallbo ÅB, Johansson RS (1984): Properties of cutaneous mechanoreceptors in the human hand related to touch sensation. *Human Neurobiol* 3:3–14

Trigeminal Response to Odors

Michael Meredith

The trigeminal nerve, or cranial nerve V, is a somatic sensory nerve, primarily sensitive to mechanical and thermal stimulation, but the branches innervating the nasal and oral cavities and the cornea also include chemosensitive fibers. Those of the nasal cavity and cornea are sensitive to airborne chemicals. They have an important function as the sensory input triggering protective reflexes—respiratory, secretory, and cardiovascular—in response to high concentrations of irritating vapors. However, they also respond to low, nonirritating concentrations of the same chemicals and to other chemicals lacking any irritant qualities. Trigeminal input can thus contribute to the sense of smell, either directly or by influencing input from olfactory (cranial nerve I) receptors.

The ophthalmic (nasociliary) division of the trigeminal supplies the cornea and the anterior nasal cavity; the nasopalatine, posterior-lateral, -medial, and -inferior nasal nerve branches of the maxillary division supply the remainder of the nasal cavity. The cell bodies of the sensory neurons lie in the gasserian ganglion, and the chemosensory terminals are free nerve endings in the respiratory epithelium and perhaps the olfactory epithelium. Where a vomeronasal organ is present the nonsensory and perhaps the sensory parts of its luminal epithelium are also innervated by the nasopalatine nerve. The oral cavity is innervated by the mandibular division of the trigeminal. Electrophysiological responses to airborne chemicals can be obtained from both ophthalmic and maxillary divisions of the nerve, and it seems likely that all branches serving the nasal cavity may carry chemoreceptive fibers.

The central projections of chemosensory fibers have not been distinguished from those of mechano- and thermosensitive fibers, but they may contribute to the uncrossed ascending projection and to the small projection reaching the solitary nucleus.

Electrophysiological recordings in tortoise suggested differences in sensitivity and selectivity between nasal trigeminal chemoreceptors and true olfactory receptors. For most substances the trigeminal receptors were less sensitive but for some appeared to be more sensitive. Clear responses in both tortoise and the rat to concentrations that would not be irritating to humans were also observed. In electrophysiological experiments, there are differences in responses to different odors at the same concentration, but it is not clear if this reflects qualitative discrimination or differences in efficiency of stimulation. Animals can give behavioral responses to airborne chemicals at high concentrations after removal of the olfactory bulbs, and patients believed to have no olfactory (cranial nerve I) function can detect a wide range of chemicals—usually with less sensitivity than normal subjects.

At irritating concentrations, trigeminal stimuli produce psychophysical sensations ranging from pungency, burning, and pain to the cooling sensation of menthol. Because these may be due to stimulation of fibers other than specific chemoreceptors including nociceptors and cold receptors, it is not clear whether odor quality can be discriminated by trigeminal chemoreceptors. All neural tissue is chemosensitive to some extent, and it is not clear whether there are trigeminal endings specialized for chemoreception or whether normal trigeminal sensation is due to the activation of other fibers. Capsiacin treatment in rats, a procedure that desensitizes pain receptors, also affects trigeminal responses. A dramatically reduced odor response and a reduction in respiratory reflexes were recorded after capsiacin treatment in rat, suggesting a major contribution to such recordings from nociceptors. Psychophysical results from Cain's laboratory point to the same conclusion for some trigeminal stimuli. Nasal pungency normally has a slow onset, and its perceived intensity depends on the total mass of chemical passed into the nose rather than its concentration. This finding is consistent with the production of tissue damage following accumulation of chemical in the mucosa rather than reversible binding to a receptor. Whatever the mechanism, it is likely that trigeminal sensory input contributes to normal olfactory sensation.

In addition to any direct contribution of trigeminal input to olfactory sensation, there is apparently a mutual inhibitory interaction between the two pathways. High concentrations of CO_2 in the nose elicit sensations of pungency, presumably by trigeminal stimulation. Such pungency adds to the olfactory sensation to increase the overall sensation magnitude when CO_2 is presented simultaneously with a nonirritating concentration of a conventional odorant (amyl butyrate). However, each substance, as its concentration is increased, produces an increasing inhibition of the characteristic sensation produced by the other substance. Neural interactions within the brain are one possible mechanism for these effects, but they could also be produced by reduction in nasal airflow to receptors caused by increased nasal secretions, mucosal vasodilation, or changes in respiration. Direct interaction between the stimulus chemicals in the nose was prevented by applying one chemical to each nostril.

Further reading

Cain WS, Murphy CL (1980): Interaction between chemoreceptive modalities of odor and irritation. *Nature* 284:255–257

Doty RL, Brugger WE, Jurs PC, Orndorff MA, Snyder PJ, Lowry LD (1978): Intranasal trigeminal stimulation from odorous volatiles: Psychometric response from anosmic and normal humans. *Physiol Behav* 20:175–185

Silver WL, Maruniak JA (1981): Trigeminal chemoreception in the nasal and oral cavities. *Chem Senses* 6:295–305

Silver WL, Mason JR, Marshall DA, Maruniak JA (1985): Rat trigeminal, olfactory and taste responses after capsiacin desensitization. *Brain Res*, 333:45–54

Tucker D (1971): Nonolfactory responses from the nasal cavity: Jacobson's organ and the trigeminal system. In: *Handbook of Sensory Physiology, IV:Pt 1, Olfaction*. Beidler LM, ed. Berlin: Springer-Verlag, pp 151–181

Vestibular System

Volker Henn

The vestibular system conveys a sense of motion and gravity. Anatomically, it consists of the labyrinths as specific receptors located in the inner ear, and the vestibular nuclei in the brain stem. Information about motion is also conveyed by the visual system and other sensory systems. The vestibular system therefore, in a sense, constitutes a subsystem of motion sense. The labyrinth consists of three semicircular canals and the otoliths (Fig. 1).

The semicircular canals are filled with fluid, the endolymph. Each canal has a cupula, which resembles a small cone consisting of hair cells whose gelatinous cover extends over the diameter of the canal (Fig. 2). During angular acceleration, i.e., rotation with varying velocity, the endolymph, because of its inertia, will generate a pressure gradient across the cupula; this leads to a bending of the hairs that in turn leads to a change of the receptor potential, i.e., the potential across the membrane of the hair cells. Because there are three canals on each side whose orientation is nearly orthogonal to each other, an acceleration in any direction can be detected as vector projection onto the plane of each canal.

The otoliths consist of two differently oriented assemblies of hair cells: the utriculus with a nearly horizontal, and the sacculus with a vertical orientation. The top of the hair cells is covered by a mass of otoconia (literally small stones made up of crystalline calcium carbonate) whose specific gravity is heavier than the surrounding tissue. Any linear acceleration leads to a shift in the position of the otoconia and thereby to a bending of the hairs extending from the underlying cells. Constant linear acceleration in an earth-vertical direction is continuously exerted by the force of gravity.

The hair cells of the cupula and of the otoliths act as mechanoelectrical transducers. A change of their generator potential when they bend leads to a specific modulation of activity in the nerve fibers contacting them. This information is carried in fibers of the vestibular nerve (i.e., the vestibular part of the eighth cranial nerve) to the vestibular nuclei. Anatomically, based on cell types and fiber connections, one can differentiate four subnuclei: the superior, medial, lateral, and inferior nucleus.

There are three main functional connections of the vestibular nuclei: a projection via the thalamus to the cortex, one to the oculomotor system, and one to the spinal cord. As is the case for all sensory systems, the cortical projection seems to be a precondition for conscious experience. The oculomotor projection leads to reflexive eye movements (vestibuloocular reflex, VOR), which serve to stabilize the retinal image in space during head movements. Any movement covering more than a 30 degree position change leads to nystagmus. The spinal projection is the anatomical basis for vestibulospinal reflexes stabilizing posture which enables us to maintain an upright stance. With the center of the body's gravity above the hip joints, we need continuous adjustment of muscle activity in the legs and trunk to keep us from falling.

An angular acceleration impulse, i.e., a sudden change in velocity of a relatively constant velocity rotation, leads to a pressure gradient across the cupula and activity in the vestibular nerve. Because of mechanical damping within the cupula-endolymph system, activity does not immediately cease after the end of acceleration, but declines with a time constant of about 6 seconds (Fig. 3). This activity is further prolonged by neuronal mechanisms in the vestibular nuclei. The end result is a time constant with values of between 10 seconds and 30 seconds. The effect is that over several seconds, activity in the vestibular nuclei becomes proportional to head velocity instead of acceleration. This mechanism of partial integration has been described as velocity storage. Thus the vestibular

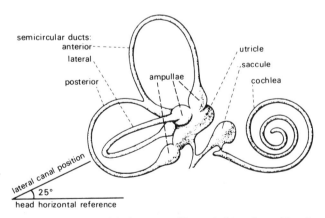

Figure 1. The labyrinth in its topographic proximity to the cochlea, the end organ for hearing. The three semicircular canals or ducts are positioned approximately at right angles to each other, the lateral one being horizontal if the head is tilted about 25 degrees, nose down. The cupula with the hair cells lies within the ampulla.

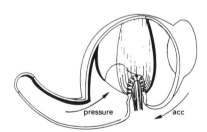

Figure 2. Schematic drawing of an ampulla. At the base are the receptor cells with hairs extending into a gelatinous mass covering the whole diameter of the ampulla.

Figure 3. Activity of a vestibular nerve fiber from the horizontal semicircular canal and nystagmus in response to horizontal angular acceleration. From above: neuronal activity, eye position, and turntable velocity. Note that neuronal activity returns to resting discharge level much faster than nystagmus declines. Recording done in a chronically prepared, alert Rhesus monkey.

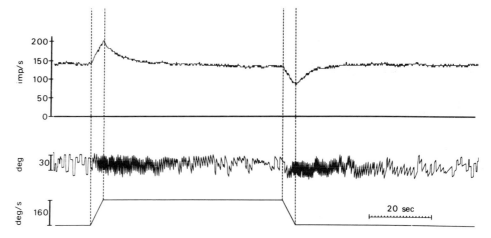

nuclei are not merely relay stations for input from the vestibular nerve but provide important information-processing functions, including multisensory interaction, as in visual-vestibular interaction and the extraction of labyrinth signals that serve to reconstruct actual head velocity.

Rotating a primate about an earth-horizontal axis (as on a barbeque spit) generates compensatory nystagmus (Fig. 4). During such constant velocity rotation in the dark, the only signals arriving at the vestibular nuclei come from the otoliths that, with each revolution, transmit a sinusoidally modulated signal according to the changing direction of the gravity vector. This signal is used to extract information about continuous rotation; consequently a continuous direction-specific nystagmus is generated. The same principle applies to nystagmus generated by somersaulting. Synaptic mechanisms for this phenomenon are unclear, but the effect is to transmit a signal about actual head velocity to the oculomotor system.

Vestibulospinal reflexes are instrumental in maintaining posture and adjusting it to changing body orientation. In humans this has been widely studied on pressure-sensitive platforms. In animals these reflexes have been investigated within the context of spinal mechanisms of motor activity.

Habituation and plasticity of the vestibular system and recovery after lesions

The multiple sensory input to the vestibular system requires continuous adjustment to generate the appropriate motor output for the oculomotor and somatic motor system. Habituation is defined as a shortening of the nystagmus time constant, which occurs with repeated low-frequency stimulation of the semicircular canals. The persistence of this effect over weeks or months differentiates it from fatigue. Plasticity refers to an altered response after the subject's exposure to a novel combination of stimuli. For instance, if a subject wears spectacles with reversing prisms, with every head movement the visual image will move in the reverse, unexpected direction. With prolonged exposure the vestibulo-ocular reflex will undergo plastic changes, and these will persist even during vestibular stimulation in darkness.

Recovery after lesions refers to changes which compensate for the initially pathological response. For instance, after one labyrinth is lesioned, animals or humans will have spontaneous nystagmus toward the contralateral side and a head tilt toward the ipsilateral side. Within days and weeks these effects sub-

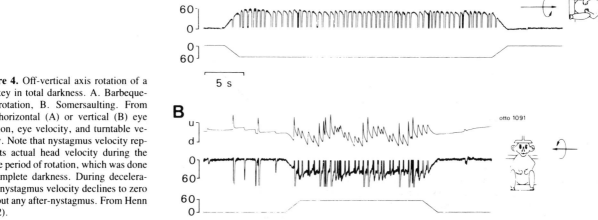

Figure 4. Off-vertical axis rotation of a monkey in total darkness. A. Barbeque-spit rotation, B. Somersaulting. From top: horizontal (A) or vertical (B) eye position, eye velocity, and turntable velocity. Note that nystagmus velocity represents actual head velocity during the whole period of rotation, which was done in complete darkness. During deceleration, nystagmus velocity declines to zero without any after-nystagmus. From Henn (1982).

side, provided the subject can actively move around. Such motor activity seems to be crucial for a recalibration of the system. If animals are immobilized, lesion effects become permanent.

Motion sickness

Motion sickness can occur during novel or unexpected combinations of various vestibular and other sensory inputs. Persons affected by it feel sick, but it is not a disease in the strict sense, because it can be induced in most persons. Many persons experience it in rough weather on a ship, when rotatory and linear accelerations combine in a nonanticipatory fashion. In space flight, the lack of gravitational force induces an activity pattern in the vestibular nerve which is different from earth and therefore unexpected for the untrained subject. Prolonged exposure always leads to adaptation to these environments, although initially subjects can become totally incapacitated. Speculation has provided no satisfactory clue to the biological "sense" or usefulness of such a response to which higher developed vertebrates (cats, dogs, monkeys) are also sensitive.

Pathology

Any noncompensated lesion of the vestibular system typically leads to vertigo and to pathological nystagmus, i.e., eye movements that do not stabilize gaze. Because of the multimodal interaction, however, even extended pathology will not lead to the loss of self-motion perception. Lesions involving the labyrinths occur in Meniere's disease (a syndrome of acute vertigo and hearing loss), and inflammatory or vascular disorders. Central lesions involving the vestibular nuclei often occur in younger patients with multiple sclerosis or in older patients with vascular diseases. Tumors and degenerative disease are other causes. Clinical tests involve the caloric irrigation of the external ear, which leads to direction-specific nystagmus. The absence of the expected response can localize a suspected lesion. To test whether gaze stabilization is normal, eye movements are measured as subjects are rotated on a turntable known as a Bárány chair.

Further reading

Cohen B, ed (1981): Vestibular and Oculomotor Physiology: International Meeting of the Bárány Society. *Ann NY Acad Sci* 374:1–892

Henn V, Cohen B, Young LR (1980): Visual-vestibular interaction in motion perception and the generation of nystagmus. *Neurosci Res Prog Bull* 18:457–651

Henn V (1982): The correlation between motion sensation, nystagmus and activity in the vestibular nerve and nuclei. In: *Nystagmus and Vertigo: Clinical Approaches to the Patient with Dizziness,* Honrubia V, Brazier M, eds. New York: Academic Press, pp 115–124

Melvill Jones G, Gonshor A (1982): Oculomotor response to rapid head oscillation (0.5–5.0 Hz) after prolonged adaptation to vision-reversal. Simple and complex effects. *Exp Brain Res* 45:45–58

Wilson VJ, Melvill Jones G (1979): *Mammalian Vestibular Physiology.* New York: Plenum

Visual-Vestibular Interaction

Volker Henn

The sense of motion uses information from different sensory receptor organs. For humans, vestibular inputs from the inner ear and the visual system are the most important. The experience of a unique sense of motion independent of the sensory system from which the information comes requires that these different messages converge. The site and neuronal mechanism of this convergence have been widely investigated in monkeys. (Anatomical and behavioral studies indicate that mechanisms in humans are similar.)

The first convergence of vestibular and visual input takes place in the vestibular nuclei, only one synapse away from the hair cells in the labyrinths (see Fig. 1). Figure 2 shows a typical cell in the vestibular nuclei with a resting discharge of 30 imp/sec which receives input from the horizontal semicircular canal. During angular acceleration in the horizontal plane, its activity increases and during constant velocity rotation, it approaches the level of spontaneous activity with a characteristic time constant (Fig. 2). If the visual surround is rotated around the stationary animal, the cell's activity increases and keeps its activity until all motion ceases. If the animal is rotated in the light, the activity induced by labyrinthine and visual stimulation combines to induce activity that, over a large range, is proportional to actual rotational velocity. This is the typical activity found in response to vestibular and visual stimuli for all cells in the vestibular nuclei. In a quantitative way there are differences in sensitivity and working range.

For the visual and labyrinthine motion information to combine, it must be coded in the same dimension. Directions are determined by the three planes of semicircular canals. As the vestibular nerve connects to three different classes of cells in the nuclei, these canal-specific planes are maintained. Activity related to visual stimulation is proportional to the velocity of the stimulus. Labyrinthine activity induced by angular acceleration is partially integrated mechanically by the cupula and neuronally in the vestibular nuclei, so that it also represents a velocity signal over a wide range of frequencies.

One can postulate that activity from the otoliths and the visual system that conveys information about linear motion combines in a similar way, but this has not been investigated in detail.

Further sites of interaction between visual and vestibular inputs are necessary because the vestibular system simultaneously serves different purposes—conveying an accurate sense of motion, moving the eyes in order to stabilize vision during head movement, and stabilizing posture. During the same head movement, we might have compensatory eye movements, or we might want to suppress them because we foveate on a moving target. The labyrinthine activity will be the same during the two movements, giving rise to the same sensation of movement. In the first case this activity will be used to induce compensatory eye movements; in the second case this activity projecting to the oculomotor system has to be canceled or suppressed. This is one of the functions of the flocculus. It has reciprocal connections with the vestibular nuclei and

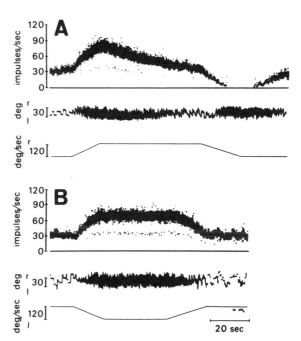

Figure 2. Activity of a neuron in the vestibular nuclei of a monkey in response to vestibular stimulation (A, angular acceleration in darkness), and visual stimulation (B, rotation of an optokinetic cylinder around the stationary animal). From the top down, instantaneous neuronal frequency, horizontal eye position, and stimulus velocity. From Waespe and Henn (1977).

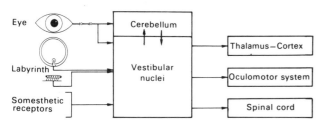

Figure 1. Diagram of how information about motion from different receptor systems converge upon the vestibular nuclei and give off three distinct pathways apart from a dense reciprocal connection with the floccular-nodular lobe of the cerebellum.

receives visual input through mossy and climbing fibers. Based on various studies, the function of the flocculus is to enhance or reduce eye velocity to adjust it to the velocity of the visual target.

In a similar interactive way inputs combine to produce somatic motor responses (mediated via the vestibulospinal tract among others) to stabilize posture.

The vestibular nuclei project to the thalamus and cortex, a necessary input for all conscious experience. As the information from individual sensory systems is combined, the unique message from the individual sense loses its identity. We cannot discriminate the individual inputs to the vestibular nuclei and so we experience but one unique sense of self-motion.

Further reading

Henn V, Cohen B, Young LR (1980): Visual-vestibular interaction in motion perception and the generation of nystagmus. *Neurosci Res Prog Bull* 18:457–651

Waespe W, Henn V (1977): Neuronal activity in the vestibular nuclei of the alert monkey during vestibular and optokinetic stimulation. *Exp Brain Res* 27:523–538

Vomeronasal Organ and Nervus Terminalis

Michael Meredith

Vomeronasal organ

General description, anatomy, and distribution. The vomeronasal organ (VNO), or Jacobson's organ, is a chemoreceptor organ, present in most tetrapod species, that is important in intraspecific chemical (pheromone) communication. The paired organs are separate from the main olfactory organ, being enclosed within the vomer bone or cartilage at each side of the base of the nasal septum in most species. The vomeronasal lumen is partially lined with chemosensory vomeronasal epithelium. This epithelium contains bipolar receptor neurons similar to those of the main olfactory epithelium except that, as a rule, they lack cilia (although cilia are reported in the dog). There is renewal of receptor cells after axotomy and degeneration and there may be continuous replacement of receptors, as in the main olfactory epithelium. The organ is present in human fetuses but in most cases is absent in adults. A large proportion (more than 50%) of clinically examined adults has vomeronasal duct openings in the nasal septum, but there are as yet no reliable reports of a functional sensory epithelium in adults. In the rat fetus, the vomeronasal nerve terminal region has a high 2-deoxyglucose uptake, possibly indicating neural activity and suggesting that the system might have some function in utero.

The organ in mammals consists of an elongated tube, opening only at the anterior end via a narrow duct into the floor of the nasal cavity (rodents, lagomorphs, and some primates) or into the nasopalatine canal (carnivores, ungulates, insectivores, and most primates). The nasopalatine canal (incisive canal, Stenson's canal) is a passageway through the palate between the nasal and oral cavities, so that in the second group of species stimuli could reach the organ via either the nose or the mouth. In some, perhaps all, mammalian species, access of stimulus molecules to the receptors is regulated by an autonomically controlled vascular pumping mechanism consisting of large blood vessels surrounding the lumen, within the vomer capsule. In some ungulates (horse, cattle, antelope) and some carnivores (cats), a behavior called "Flehmen" or "testing," typically seen when males investigate female secretions, has been suggested as an outward manifestation of active vomeronasal stimulation. The stereotyped posture involved (lifted head, curled back upper lip) is said to open the narrow passageways to the receptors.

The VNO is prominent in reptiles (except crocodilians), especially in snakes, where the openings are into the roof of the mouth and the organs are stimulated by materials delivered to their ducts by the flicking of the bifid tongue. The VNO is used for prey trailing and capture as well as intraspecific chemical communication (courtship, mating, and aggregation). Amphibians have the organ but its function in this group is not known. Cetacean marine mammals, birds, and fish lack the organ.

Function. Most of the work on vomeronasal function has been in rodents and snakes, and most has involved lesions of the vomeronasal or olfactory systems in order to reveal deficits in behavior or physiological function. In rodents, intraspecific chemical communication (pheromone communication) can produce dramatic effects on reproductive behavior and physiology—many of which depend on chemosensory input from the vomeronasal system. For example, female mice housed in groups produce a urinary chemosignal that suppresses estrus in other females. Male mice and other rodents produce a urinary chemosignal that accelerates puberty in immature females of the same species. In both cases, removal of the VNO prevents the response. Such effects are probably due to an influence of vomeronasal input on hormone levels. Both sexes of many species show changes in hormone levels, principally luteinizing hormone releasing hormone (LHRH), LH, and prolactin in response to chemical signals from opposite-sex individuals, and these hormonal responses may be responsible for the physiological changes. In mice and hamsters, removal of the vomeronasal organs prevents the hormonal changes normally observed after exposure of males to female chemosignals. Behavior is also dependent on chemosensory input. Male mice and hamsters do not mate if they are surgically deprived of both olfactory and vomeronasal sensory input. In sexually experienced adults, either olfactory or vomeronasal input is sufficient to allow courtship (e.g., ultrasonic calling) and mating in response to female chemical cues, but in sexually naive animals the removal of the vomeronasal organs produces much greater deficits in these behaviors than does removal of the olfactory input alone. Thus the vomeronasal input may be responsible for eliciting preprogrammed behavior (e.g., mating), but this behavior may also be elicited by olfactory or other sensory input once a period of association between vomeronasal and olfactory input has occurred.

Central neural connections. The central neural connections of the vomeronasal system are consistent with its proposed role in intiating social and reproductive responses. The vomeronasal receptor axons project from the epithelium where the cell bodies are located, to the accessory olfactory bulb (AOB)—a separate part of the olfactory bulb usually posterior-dorsal to the main olfactory bulb. Second-order accessory olfactory cells in the AOB project to the corticomedial amygdala which, in those species with a VNO, apparently receives no direct projections from the main olfactory bulb although there are association fiber inputs from other olfactory projection areas. Thus the central vomeronasal projections are separate from the main olfactory projections as far as the amygdala. Cells in the corticomedial amygdala project to preoptic and hypothalamic areas concerned with hormonal control, reproduction, and vegetative function including feeding. There is apparently no neocortical projection of the vomeronasal system, and it has been suggested that vomeronasal sensory input may be unavailable to cognitive processes. In humans the corticomedial amygdala receives main olfactory bulb input but may have evolved as an accessory olfactory projection area. In snakes the AOB projects to the nucleus sphericus, supposedly an amygdaloid structure, and is again separate from the main olfactory input to this level.

Sensitivity and Selectivity. Electrophysiological recordings from the vomeronasal system in the turtle showed responses to volatile substances that also stimulated the main olfactory receptors, although some slight differences in sensitivity were apparent. The vomeronasal system appeared to be more sensitive to some odor chemicals and less sensitive to others. A few recordings from mammals also suggest general rather than highly selective sensitivity. However, identified substances known to be involved in the behavioral and physiological responses to vomeronasal input have not been available for testing. A more selective response of the vomeronasal receptors to such substances at very low concentrations, analagous to the selectivity of specialist pheromone receptors in insects, cannot yet be ruled out. It is not yet clear whether the system can discriminate qualitative differences between stimuli or whether its input simply reflects the intensity of stimulation regardless of the stimulating chemical. In both mammals and reptiles there is evidence that the VNO is involved in detecting large nonvolatile components of natural secretions. In some cases, chemical analysis of odorous materials involved in chemical communication has implicated large molecules (possibly proteins) as chemical signals in behaviors where there is also evidence for vomeronasal involvement. Nonvolatile tracer substances mixed with appropriate natural stimulus materials can be observed inside the vomeronasal lumen, in various species, after the animal has been allowed to investigate and contact the stimulus. However, direct electrophysiological demonstration of vomeronasal response to nonvolatile substances has not been tested and the possibility that nonvolatiles may also stimulate the main olfactory receptors should not be ignored.

Nervus terminalis

Most of the evidence implicating vomeronasal input in social and sexual behavior and physiology relies on selective lesions of some part of the system and the subsequent observation of deficits in behavioral or hormonal response. There is a remote possibility that some or all of the deficits observed are due to damage to an alternative neural system, the *nervus terminalis* (nt), which might be damaged in each case of intended vomeronasal lesion. The nt is ubiquitous in vertebrates, including humans. It innervates the olfactory and vomeronasal cavities and projects directly to the medial septum/preoptic area (and to the retina in some fish). There may be both sensory and autonomic components. The nerve and its ganglion contain high levels of LHRH-like immunoreactivity, and it has been implicated in pheromonal responses in fish, although its chemosensitivity has not been demonstrated as yet.

Further reading

Demski LS, Northcutt RG (1983): The terminal nerve: A new chemosensory system in vertebrates? *Science* 220:435–437

Halpern M (1983): Nasal chemical senses in snakes. In: *Advances in Vertebrate Neuroethology,* Ewert JP, Capranica R, Ingle DJ. New York: Plenum Press, pp 141–176

Meredith M (1983): Sensory physiology of pheromone communication. In: *Pheromones and Reproduction in Mammals.* Vandenbergh JK, ed. New York: Academic Press, pp 199–252

Meredith M (1983): Vomeronasal lesions before sexual experience impair male mating behavior in hamsters. *Proc Int Union Physiol Sci* 15:368

Tucker D (1971): Nonolfactory responses from the nasal cavity: Jacobson's organ and the trigeminal system. In: *Handbook of Sensory Physiology IV Pt 1 Olfaction,* Beidler LM, ed. Berlin: Springer-Verlag, pp 151–181

Winans SS, Lehman MN, Powers JB (1982): Vomeronasal and olfactory CNS pathways which control hamster mating behavior. In: *Olfaction and Endocrine Regulation,* Breipohl W, ed. London: IRL Press, pp 23–34

Wysocki CJ (1982): The vomeronasal organ: Its influence upon reproductive behavior and underlying endocrine systems. In: *Olfaction and Endocrine Regulation.* Briepohl W, ed. London: IRL Press, pp 195–208

Wysocki CJ (1979): Neurobehavioral evidence for the involvement of the vomeronasal system in mammalian reproduction. *Neurosci Biobehav Rev* 3:301–341

Contributors

Linda M. Bartoshuk   Department of Psychology, John B. Pierce Foundation Laboratory, New Haven, Connecticut 06519, U.S.A.

Lloyd M. Beidler   Department of Biological Science, Florida State University, Tallahassee, Florida 32306-3050, U.S.A.

Michael V.L. Bennett   Professor and Chairman of Neuroscience, Albert Einstein College of Medicine, Bronx, New York 10461, U.S.A.

John F. Brugge   Department of Neurophysiology and Waisman Center on Mental Retardation and Human Development, University of Wisconsin, Madison, Wisconsin 53705, U.S.A.

Theodore H. Bullock   Department of Neurosciences, School of Medicine, University of California, San Diego, La Jolla, California 92093, U.S.A.

Paul R. Burgess   Department of Physiology, University of Utah School of Medicine, Salt Lake City, Utah 84108, U.S.A.

Michel Cabanac   Université Claude Bernard, Laboratoire de Physiologie, Faculté de Médicine Lyon-Sud, F-69600 Oullins, France

William S. Cain   Department of Psychology, John B. Pierce Foundation Laboratory, New Haven, Connecticut 06519, U.S.A.

Robert R. Capranica   Section of Neurobiology and Behavior, Cornell University, Ithaca, New York 14853, U.S.A.

W. Crawford Clark   New York State Psychiatric Institute and Department of Psychiatry, College of Physicians & Surgeons, Columbia University, New York, New York 10032, U.S.A.

Richard E. Coggeshall   Marine Biomedical Institute, University of Texas Medical Branch, Galveston, Texas 77550-2772, U.S.A.

Trygg Engen   Walter S. Hunter Laboratory of Psychology, Brown University, Providence, Rhode Island 02912, U.S.A.

Alan N. Epstein   Professor of Behavioral Neuroscience, Department of Biology, Joseph Leidy Laboratory, University of Pennsylvania, Philadelphia, Pennsylvania 19104, U.S.A.

Howard H. Erickson   Professor of Physiology, Department of Anatomy and Physiology, College of Veterinary Medicine, Kansas State University, Manhattan, Kansas 66506, U.S.A.

Albert I. Farbman   Department of Neurobiology and Physiology, Northwestern University, Evanston, Illinois 60201, U.S.A.

Tsuneo Fujita   Professor and Chairman, Department of Anatomy, Niigata University School of Medicine, Asahi-Machi, Niigata 951, Japan

Peter Görner   Department of Biology, University of Bielefeld, D-4800 Bielefeld 1, Federal Republic of Germany

Ashton Graybiel   Naval Aerospace Medical Research Laboratory, Naval Air Station, Pensacola, Florida 32508, U.S.A.

G.A. Monti Graziadei   Department of Biological Science, The Florida State University, Tallahassee, Florida 32306, U.S.A.

P.P.C. Graziadei   Department of Biological Science, The Florida State University, Tallahassee, Florida 32306, U.S.A.

Peter H. Hartline   Eye Research Institute of Retina Foundation, Boston, Massachusetts 02114, U.S.A.

Gerald L. Hazelbauer   Biochemistry/Biophysics Program, Washington State University, Pullman, Washington 99164-4660, U.S.A.

Robert I. Henkin   Center for Molecular and Sensory Disorders, Georgetown University Medical Center, 5125 MacArthur Blvd, Washington, District of Columbia 20016, U.S.A.

Volker Henn   Professor of Neurology, Neurology Department, University Hospital, CH-8081 Zürich, Switzerland

Kenneth W. Horch   Department of Physiology, University of Utah School of Medicine, Salt Lake City, Utah 94108, U.S.A.

Franz Huber   Scientific Member, Max-Planck Society, Director, Institute for Behavioral Physiology, D-8131 Seewiesen, Federal Republic of Germany

A.J. Hudspeth   Department of Physiology, University of California, School of Medicine, San Francisco, California 94143, U.S.A.

Ainsley Iggo   Department of Veterinary Physiology, Royal (Dick) School of Veterinary Studies, University of Edinburgh, Summerhall, Edinburgh, EH9 1QH, Scotland

Jon H. Kaas   Department of Psychology, Vanderbilt University, Nashville, Tennessee 37240, U.S.A.

Eric Barrington Keverne   Department of Anatomy, University of Cambridge, Cambridge CB2 3DY, England

Joel D. Knispel   Coordinator of Psychology and Music Studies, Director of Psychoacoustic Research, Peabody Conservatory of Music and Department of Psychology, The Johns Hopkins University, Baltimore, Maryland 21202, U.S.A.

Masakazu Konishi   Division of Biology 216–76, California Institute of Technology, Pasadena, California 91125, U.S.A.

Lawrence Kruger   Department of Anatomy, UCLA School of Medicine, Center for the Health Sciences, University of California, Los Angeles, California 90024, U.S.A.

James M. Lipton   Professor, Physiology and Anaesthesiology Departments, Southwestern Medical School, University of Texas Health Science Center at Dallas, Dallas, Texas 75235, U.S.A.

Gerald E. Loeb   Laboratory of Neural Control, NINCDS, National Institutes of Health, Bethesda, Maryland 20892, U.S.A.

Donald M. MacKay   (Died, February 6, 1987) Emeritus Professor, Department of Communication and Neuroscience, University of Keele, Keele, Staffordshire ST5 5BG, England

Peter B.C. Matthews   University Laboratory of Physiology, Oxford, OX1 3PT, England

D. Ian McCloskey   School of Physiology and Pharmacology, University of New South Wales, Kensington, Sydney, Australia

Michael Meredith   Department of Biological Science, Florida State University, Tallahassee, Florida 32306, U.S.A.

Björn Meyerson   Department of Neurosurgery, Karolinska Sjukhuset, S-10401 Stockholm, Sweden

Maxwell M. Mozell   Professor of Physiology, Department of Physiology, State University of New York Health Science Center at Syracuse, Syracuse, New York 13210, U. S.A.

Edward Perl   Department of Physiology, University of North Carolina at Chapel Hill, Chapel Hill, North Carolina 27514, U.S.A.

Charles E. Poletti   Associate Professor of Neurosurgery, Massachusetts General Hospital, Harvard Medical School, Boston, Massachusetts 02114, U.S.A.

Ernst Pöppel   Institut für Medizinische Psychologie, Ludwig-Maximilians-Universität, D-8000 München 2, Federal Republic of Germany

F. Clifford Rose   Director, Academic Unit of Neuroscience, Charing Cross & Westminister Medical School, London W68RF, England

James D. Rose   Department of Psychology, University of Wyoming, Laramie, Wyoming 82071, U.S.A.

Jerome N. Sanes   Human Motor Control Section, Medical Neurology Branch, National Institute of Neurological and Communicative Disorders and Stroke, NINCDS, Bethesda, Maryland 20892, U.S.A.

Dietrich Schneider   Max-Planck-Institut für Verhaltensphysiologie, D-8131 Seewiesen/Starnberg, Federal Republic of Germany

Gordon M. Shepherd   Section of Neuroanatomy, Yale University School of Medicine, New Haven, Connecticut 06510, U.S.A.

David C. Spray   Department of Neuroscience, Albert Einstein College of Medicine, Bronx, New York 10461, U.S.A.

Douglas G. Stuart   Department of Physiology, College of Medicine, University of Arizona Health Sciences Center, Tucson, Arizona 85724, U.S.A.

Nobuo Suga   Professor of Biology, Department of Biology, Washington University, St. Louis, Missouri 63130, U.S.A.

William H. Sweet   Professor of Surgery, Emeritus, Harvard Medical School, Senior Neurosurgeon, Neurosurgical Service, Massachusetts General Hospital, Boston, Massachusetts 02114, U.S.A.

Ronald R. Tasker   Division of Neurosurgery, Toronto General Hospital, 200 Elizabeth St., Toronto, Ontario M5G 2C4, Canada

Lars Terenius   Institutionen für Farmakologi, Uppsala Universitets, Biomedicum, Box 591, Uppsala, Sweden

John Thorson   Consultant, Max-Planck-Institut für Verhaltensphysiologie, D-8131 Seewiesen, Federal Republic of Germany

Robert P. Tuckett   Department of Physiology, University of Utah School of Medicine, Salt Lake City, Utah 84108, U.S.A.

A.B. Vallbo   Professor, Department of Physiology, Umeå University, S-901 87 Umeå, Sweden

Charles Walcott   Laboratory of Ornithology and Section of Neurobiology and Behavior, Cornell University, Ithaca, New York 14850, U.S.A.

Patrick D. Wall   Director, Cerebral Functions Group, Department of Anatomy and Embryology, University College London, Gower Street, London WCIE 6BT, England

Thomas A. Woolsey   James L. O'Leary Division of Experimental Neurology & Neurosurgery, Washington University School of Medicine, Saint Louis, Missouri 63110, U.S.A.

Laurence R. Young   Man-Vehicle Laboratory, Department of Aeronautics and Astronautics, Massachusetts Institute of Technology, Cambridge, Massachusetts 02139, U.S.A.